Formulation Simplified

Formulation Simplified

Finding the Sweet Spot through Design and Analysis of Experiments with Mixtures

Mark J. Anderson
Patrick J. Whitcomb
Martin A. Bezener

A PRODUCTIVITY PRESS BOOK

Routledge
Taylor & Francis Group
711 Third Avenue, New York, NY 10017

Productivity Press is an imprint of Taylor & Francis Group, an Informa business

Printed on acid-free paper

International Standard Book Number-13: 978-1-1380-5604-6 (Hardback)
International Standard Book Number-13: 978-1-1380-5601-5 (Paperback)
International Standard Book Number-13: 978-1-3151-6557-8 (eBook)

Visit the Taylor & Francis Web site at
http://www.taylorandfrancis.com

and the Productivity Press site at
www.productivitypress.com

Contents

Preface

All that is gold does not glitter; Not all those who wander are lost.

—J. R. R. Tolkien (*The Fellowship of the Ring*)

This book rounds out our series of "Simplified" books (Anderson and Whitcomb, 2015, 2016) into a trilogy on the design of experiments (DOE). It may not achieve the stature of Tolkien's towering trio—*The Lord of the Rings*, but the detailing of mixture design completes our quest to provide the statistical tools needed by modern-day industrial experimenters. The beneficiaries of this third "Simplified" book will be formulators of alloys, beverages, chemicals, cosmetics, construction materials (such as concrete), food, flavors, pharmaceuticals, paints, plastics, pulp, paper rubber, textiles, and so forth, that is, any product made from stuff.

Formulation Simplified is derived from a popular workshop on mixture design that my coauthor, Pat, developed over twenty years ago. He's worked unstintingly to continuously incorporate new statistical methods that prove to be of practical use. More recently, statistician, Martin Bezener, joined our team at Stat-Ease and took to mixtures like a barista to coffee. However, it's one thing to be trained intensively by expert instructors like Pat or Martin, but another thing to learn on your own from a book. That's where I come in by making these powerful statistical tools of experimental design and analysis as unintimidating as possible in a self-study, written format. Luckily, I can rely on Pat and Martin to bolster any inadequate mathematical details, thus helping us maintain statistical rigor throughout. If we feel that this may create too much information (TMI) for some readers, the in-depth explanations go into sidebars or appendices that can be glossed over (at least on the first go-through!).

What differentiates *Formulation Simplified* from the standard statistical texts on mixture design by Cornell (2002) and Smith (2005) is that we make things relatively easy and fun to read. To convey my experience that "experimental design can lead to feelings of pure joy, especially when it points in the direction of improvement" (Lynne Hare), I provide a whole new collection of delightful original studies that illustrate the essentials of mixture design and analysis. Solid industrial examples are offered as problems at the end of many chapters for those who are serious about trying new tools on their own. Statistical software to do the computations can be freely accessed via a website developed in support of this book. There you will also find answers posted. See *About the Software* for instructions.

Early in my career as a chemical engineer working in process development, I discovered that many of the chemists in our research and development center—especially those most brilliant in their field—failed to appreciate the power of planned experimentation. Furthermore, they disliked the mathematical aspects of statistical analysis. To top off their DOE phobia, these otherwise competent chemists also dismissed predictive models based only on empirical data. Ironically, in the hands of subject matter experts like these elite chemists, the statistical methods of mixture design and analysis provide the means for rapidly converging optimal compositions—the "sweet spot," as I like to call it. That is the true gold that awaits those who are willing to take a new path.

Mark J. Anderson

Acknowledgments

Pat, Martin, and I are indebted to several contributors to the development of mixture-design methods, particularly, Wendell Smith and Greg Piepel. However, we must especially acknowledge John Cornell. His landmark book, *Experiments with Mixtures* (Wiley, 1981) served as our template. With John's passing in 2016, our community of industrial statisticians lost a bright light—a fellow with unbounded enthusiasm for the design of experiments (DOE) geared to formulators. We dedicate this book to John with the hope that it carries his torch for mixture design forward.

Mark J. Anderson

Authors

Mark J. Anderson, PE, CQE, MBA, is a principal and general manager of Stat-Ease, Inc. (Minneapolis, Minnesota). He is a chemical engineer by profession, who also has a diverse array of experience in process development (earning a patent), quality assurance, marketing, purchasing, and general management. Before joining Stat-Ease, he spearheaded an award-winning quality improvement program (generating millions of dollars in profit for an international manufacturer) and served as general manager for a medical device manufacturer. His other achievements include an extensive portfolio of published articles on the design of experiments (DOE). Anderson authored (with Whitcomb) *DOE Simplified: Practical Tools for Effective Experimentation, 3rd Edition* (Productivity Press, 2015) and *RSM Simplified: Optimizing Processes Using Response Surface Methods for Design of Experiments, 2nd Edition* (Productivity Press, 2016).

Patrick J. Whitcomb, MS, is the founding principal of Stat-Ease, Inc. Before starting his own business, he worked as a chemical engineer, quality assurance manager, and plant manager. Whitcomb developed Design-Ease® software, an easy-to-use program for design of two-level and general factorial experiments, and Design-Expert® software, an advanced user's program for response surface, mixture, and combined designs. He has provided consulting and training on the application of design of experiments (DOE) and other statistical methods for decades. In 2013, the Minnesota Federation of Engineering, Science, and Technology Societies (MFESTS) awarded Whitcomb the Charles W. Britzius Distinguished Engineer Award for his lifetime achievements.

Martin A. Bezener, PhD, is a principal and statistician with Stat-Ease, Inc. He did his graduate studies at the University of Minnesota, Twin Cities. There he spent a year at the Statistical Consulting Center of the School of Statistics working on a wide variety of projects with university researchers. He also taught undergraduate-level statistics for several years. In addition to his role as a consultant on DOE, Martin takes point at Stat-Ease for researching new methodology and developing algorithms for coding into the publisher's software.

Introduction

> There are many paths to enlightenment. Be sure to take one with a heart.
>
> **—Lao Tzu**

To avoid disheartenment, this book swoops down from above the forest to treetop level, where it remains for the most part. Go ahead and enjoy the ride for the first pass through *Formulation Simplified*—it will be a far easier read than any other statistical textbook you are likely to see, especially, if you skip the formula-laden appendices and the serious sidebars (read the trivial ones just for fun).

A MODEL FOR PREDICTING WHAT FORMULAS DO TO GENERAL READERSHIP

Mathematics Professor Roger Penrose prefaces his book The Emperor's New Mind *(Oxford University Press, 1989) with this note to readers (excerpted): "At a number of places I have resorted to the use of mathematical formulae,...unheeding of warnings...that each such formula will cut the general readership by half...I recommend [you] ignore that line completely and skip over to the next actual line of text!...If armed with new confidence, one may return to that neglected formula and try to pick out some salient features. The text itself may be helpful in letting one know what is important and what can be safely ignored about it."*

Great advice!

(Kudos to colleague Neal Vaughn for alerting us about this rule of thumb!)

However, if you genuinely hope to master the tools for design and analysis of experiments with mixtures go back and do your homework via the practice problems. Be sure to pursue the links to web-based content that provides many of the details you will need to interpret the statistical analyses and graphics from the software we make available or others that offer the same features (there are several good alternatives that can be easily searched out if not already at your fingertips from your enterprise's server).

Finally, to leave no leaf unturned, consider going back to the first two books in this trilogy—*DOE Simplified* and *RSM Simplified*. Even if you have already read these two books, it will be useful to leaf through them (pun intended) and review the detailed statistical tools presented that remain useful for mixture design and analysis, for example—diagnostics for model validation.

See the flowchart for specific parts of the prior Simplified books that will be very relevant to what's covered in *Formulation Simplified*.

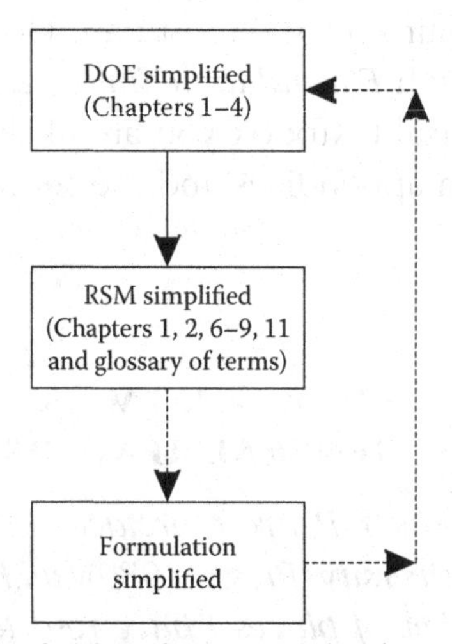

Flowchart of recommended readings

Ultimately this all becomes just an academic exercise (like reading a textbook about how to ride a bicycle) unless you actually take these tools for a spin on your own. It will not be hard to find a proper application for mixture design and analysis—just consider your favorite food or beverage and search out the sweet spot in their formulation. You will see plenty of other ideas throughout the book on experiments you can do at home, but better yet, dive in on something that will benefit your sponsor or employer.

The secret weapon you will develop by reading this book and putting its tools into practice is the ability to handle many ingredients simultaneously—not just one at a time, as dictated from time immemorial by "the scientific method." As you will see from example after example, one needn't hold all else constant while changing only one thing. Instead, take advantage of modern-day parallel processing schemes with mixture designs that provide multicomponent testing. This is the forest we hope you will see from the highest level, which then will provide the necessary motivation for sharpening up your ax before going back to hacking at the trees of formulation development.

Chapter 1

Getting Your Toe into Mixtures

Simplicity is the ultimate sophistication.

—Leonardo da Vinci

Come on in—the water's fine! Ok, maybe you'd do best by first sampling the temperature of the pool with just your finger or toe. That's what we will try to do in this chapter—start with the simple stuff before getting too mathematical and statistical about mixture design and modeling for optimal formulation. Our two previous books, *DOE Simplified* and *RSM Simplified*, both featured chapters on mixture design that differentiate this tool from factorials and response surface methods (RSM), respectively. However, if you did not read these books, that's OK. We will start with an empty pool and fill it up for you!

It's natural to think of mixtures as liquids, such as the composition of chemicals a pool owner must monitor carefully to keep it sanitary. However, mixtures can be solids too, such as cement or pharmaceutical excipients—substances that bind active ingredients into pills. The following two definitions of mixtures leave the form of matter open:

- "Mixtures are combinations of ingredients (components) that together produce an end product having one or more properties of interest"—Cornell and Piepel (2008).

- "What makes a mixture?
  1. The factors are ingredients of a mixture.
  2. The response is a function of proportions, not amounts.
  - Given these two conditions, fixing the total (an equality constraint) facilitates modeling of the response as a function of component proportions"—Whitcomb (2009).

The first definition by Cornell and Piepel provides a practical focus on products and the interest that formulators will naturally develop for certain properties of their mixture (as demanded by their clients!). However, the second specification of a mixture from Pat presented more concise conditions that provide a better operational definition. He suggests that formulators ask themselves an easy question: "If I double everything, will I get a different result?" If the answer is no, such as it would be for a sip of sangria from the glass versus the carafe, for example (strictly for the purpose of tasting!), then mixture design will be the best approach to experimentation.

THE QUINTESSENTIALS OF MIXTURES

Mixture experiments date back to ancient times when it was thought that all matter came from four elements: water, earth, fire, and air. For centuries, alchemists sought the magical fifth element, called the "quintessence," which would convert base metals to gold. Petrochemicals made from "black gold" (oil) became the focus of Henri Scheffé's pioneering article in the field of statistical design and analysis of mixtures—Experiments with Mixtures (1958). Perhaps this interest was sparked by his work in World War II, which remains obscure (not revealed at the time) but can be described in general terms as having to do with "the effects of impact and explosion" (Daniel and Lehmann, 1979). Thus one assumes that Scheffé studied highly exothermic mixtures in his formative years as an industrial statistician.

Later, Sahrmann et al. (1987) created a stir in the field for their mixture experiments on Harvey Wallbanger's popular cocktail in the 1970s. A potential problem with mixture experiments on alcoholic drinks is that, unless the tasters are professional enough to refrain from drinking

(*Continued*)

the little they sip, after several samples, the amount of alcohol ingested could matter—not just the proportions. Therefore, one should never permit sensory evaluators to consume alcoholic beverages—only sip, spit and rinse mouth afterward with water. Keep that in mind if you wish to apply the methods of Cornell's landmark book (2002) to such purpose (e.g., if you become inspired by our case study in Chapter 2 on blending beers).

PS: Of course, some mixtures are better liberally applied—for example, primer paint—the more, the better for hiding power. This would be a good candidate for a "mixture-amount" design of formulation experiment. They require more complicated approaches and modeling so let's set this aside for now. We will devote our full attention to "mixture-amount" experiments towards the end of the book.

> With mixtures, the property studied depends on the proportions of the components present, but not on the amount.
>
> **—Henri Scheffé**

All That Glitters Is Not Gold

Let's now dive in on the shallow end of design and analysis of experiments with mixtures. The easiest approach will be to lead by example via a case study.

Some years ago, Mark enjoyed a wonderful exhibit on ancient gold at the Dallas Art Museum. It explained how goldsmiths adulterated gold with small amounts of copper to create a lower melt-point solder that allowed them to connect intricately designed filigrees to the backbone of bracelets and necklaces. This seemed very mysterious, given that copper actually melts at a higher temperature than gold! However, when mixed, these two metals melt at a lower temperature than either one alone. This is a very compelling example of synergism—a surprisingly beneficial combination of ingredients that one could never predict until they are actually mixed for experimental purposes (Figure 1.1).

Figure 1.1 This exquisite necklace, now in London's British Museum, came from a necropolis (burial site) on Rhodes. It features Artemis, the Greek goddess of hunting. (Courtesy of Bridgeman Art Library, London/New York.)

A EUREKA MOMENT!

You may recall from studying Archimedes' principle of buoyancy in which this Greek mathematician, physicist, and inventor who lived from 287–212 BC was asked by his King (Hiero of Syracuse) to determine whether a crown was pure gold or was alloyed with a cheaper, lighter metal. Archimedes was confused as to how to prove this, until one day, when he started observing the overflow of water from his bathtub, he suddenly realized that, since gold is denser, a given weight of gold represents a smaller volume than an equal weight of the cheap alloy. Therefore, a given weight of gold would displace less water. Delighted at his discovery, Archimedes ran home without his clothes, shouting "Eureka! Eureka!" which means "I have found it! I have found it!" When you make your discovery with the aid of mixture design for an optimal formulation, feel free to yell "Eureka!" as well, but wait until you get dressed.

PS: If you have a copy of *DOE Simplified, 3rd Edition*, see the Chapter 9 sidebar "Worth its weight in gold?" It provides information on a linear blending model we derived based on the individual densities of copper versus the much heavier (nearly double) gold.

The ancient Greek and Roman goldsmiths mixed their solder by a simple recipe of 2 parts gold and 1 part copper (Humphrey et al., 1998). The use of "parts," while extremely convenient for formulators as a unit of measure is very unwieldy for doing mathematical modeling of product performance. The reason is obvious, the more parts of one material that you add, the more diluted the other ingredients become, but there is no quantitative accounting for this. For example, some goldsmiths added 1 part of silver to the original recipe. That now brings the total to 4 parts, and thus the gold becomes diluted further (2 parts out of 4% or 50%, versus the original concentration of 2/3 or about 67%). Therefore, one of the first things we must do is wean formulators wanting to use modern tools of mixture design off the old-fashioned use of parts. In this case, it will be convenient to specify the metal mixture by weight fraction—scaled from zero (0) to one (1). However, all that matters is that the total is fixed, such as one for the weight fraction. Alternatively, if our goldsmith used a 50-milliliter crucible, then the ingredients could be specified by volume—provided that when added together, they will always be 50 mL. You will see various units of measurements used in mixture designs throughout this book, although the most common may be by weight. Regardless, the first thing we will always specify is the total.

Getting back to the task at hand, let's see the results for the temperature at which various mixtures of gold and copper begin to melt. Assume this was done in ancient times when measurements were not very accurate. (This is a pretend experiment!) We've covered the entire range from zero to one of each metal (Table 1.1).

Notice that the table sorts the blends by their purity of gold. The actual order for experimentation can be assumed to be random. As emphasized in both our previous books on statistical design, randomization provides insurance against lurking variables such as warm-up effects from the furnace, cross-contamination in the crucible, learning curves of operators and so forth. As the inventor of modern-day industrial statistics, R. A. Fisher said, "Designing an experiment is like gambling with the devil: only a random strategy can defeat all his betting systems."

Table 1.1 Melt Points of Copper versus Gold and Mixtures of the Two

Blend #	*Point Type*	*Blend Type*	*Gold (wt fraction)*	*Copper (wt fraction)*	*Melt Point (Deg C)*
1	Vertex	Pure	0.00	1.00	1073
2	"	"	0.00	1.00	1063
3	"	"	0.00	1.00	1083
4	Axial check blend	Quarter	0.25	0.75	955
5	Third edge	Third	0.33	0.67	951
6	Centroid	Binary	0.50	0.50	926
7	Third edge	Third	0.67	0.33	929
8	Axial check blend	Quarter	0.75	0.25	952
9	Vertex	Pure	1.00	0.00	1049
10	"	"	1.00	0.00	1036

PURE BLENDS REPLICATED TO PROVIDE PURE ERROR

To provide an estimate of pure error, the "pure blends" (yes, that is an oxymoron) are replicated several times, with one more for copper (blend #s 1–3) than for gold (#s 9–10), thus saving on the precious metal. These replicates provide three degrees of freedom (df), two from the copper (df = 3 − 1) and one from the gold (df = 2 − 1). As a benchmark for testing lack of fit (see Appendix 1B for details on this statistic), this is a marginal number of degrees of freedom (df) for pure error. If gold were not so costly, one more replicate would be better. That increases the df for pure error to four, which we recommend at a minimum.

Another important element of this experimental design is the replication designated in the descriptor columns (point type and blend type) by ditto marks ("). We advise that at least three blends be replicated in the randomized plan, preferably four or more. These provide a measure of pure error, desirable for statistical purposes, but as a practical matter, the replicates offer an easy way for formulators to get a feel for their inevitable variations in blending the materials and measuring the response(s)—simply look at the results from run-to-run made by the same recipe.

Generating a Beautiful Response Surface—Like a String of Rubies on a Gold Strand!

Ok, perhaps we are getting carried away in our enthusiasm for using data from a well-designed mixture experiment to produce a very useful plot of predicted response at any given composition. Here is our equation, fitted from the experimental data by least squares regression, for modeling the melt point (m.p.) as a function of the two ingredients: gold and copper, symbolized by x_1 and x_2; respectively. These input values are expressed on a coded scale of zero to one, which statisticians prefer for modeling mixtures.

$$\text{Melt point} = 1043\,x_1 + 1072\,x_2 - 536\,x_1x_2$$

This mixture model, developed by Henri Scheffé (1958), is derived from the conventional second-order polynomial for process RSM, called a quadratic equation. The mathematical details are spelled out by Cornell (2002). Two things distinguish Scheffé's polynomial from that used for RSM. First, there is no intercept. Normally this term represents the response when factors are set to zero—set by standard coding to their midpoints for process modeling. However, a mixture would disappear entirely if all the components went to zero—we can't have that! The second aspect of this second-order mixture-model that differs from those used for RSM is that it lacks the squared terms. Again, refer to Cornell's book for the mathematical explanation, but suffice it to say for our purposes that the x_1 x_2 terms capture the nonlinear blending behavior—in this case, one that is synergistic, that is—a desirable combination of two components.

THE JARGON OF DESIGN OF EXPERIMENTS ON MIXTURES VERSUS PROCESS

Cornell and other experts are very particular on how one describes the elements of design and analysis for mixture experiments. For example, always refer to the manipulated variables as "components"—not factors. Those of you who are familiar with factorial DOE and RSM will see other aspects that closely parallel in this book on mixture design but which are named differently. One of the traps you may fall into is referring to the second-order mixture-term x_ix_j as an interaction. If you say this in the presence of the real experts, you'd best duck and cover as school children were advised

(Continued)

in the Cuban Missile Crisis (dating ourselves here!)—the proper descriptor is "nonlinear blending." Although it seems picky, there is a good reason for this: Curvature and interaction terms that appear in process models become partially confounded due to the mixture constraint that all components sum to a fixed total. Do not fight this—just don't say these words! To develop a high level of expertise in any technical field, one must learn the technical terms and express them with great care to maintain precision in communication. This strategy can be a pain, but it provides great gain.

Observe that although this experiment requires the control of two inputs—gold versus copper, only one X-axis is needed on the response surface plot shown in Figure 1.2. That is because of the complete inverse correlation of one component with the other—as one goes up the other goes down and vice versa. In statistical terms, this can be expressed as $r = -1$, where r symbolizes correlation, and the minus sign indicates the inverse relationship.

Let's see how that model for m.p. connects to the graph. First, the coefficient of 1043 for x_1 estimates that temperature in degrees Celsius at which pure gold melts. On the other hand, pure copper melts at a higher temperature—estimated from this experiment to be 1072°C. Always keep in mind that results will vary from any given experiment, which represents only a sampling of the true population of all possible results from your process—an unknown and unknowable value. The predicted values represented by

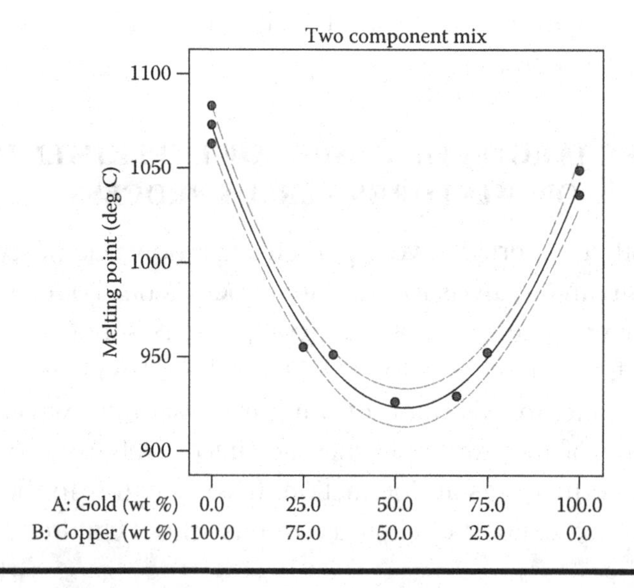

Figure 1.2 Response surface for melt point of copper versus gold and their mixtures.

the solid line in Figure 1.2 are simply an estimate. This is accentuated by the addition of 95% confidence bands (dashed) to the plot. What really counts is that, as a practical matter, the predictions serve the purpose of the goldsmith for using copper to formulate an optimal jewelry-solder.

The most intriguing feature of this mixture model is the large negative coefficient of 536 on the x_1x_2 terms. The analysis of variance (ANOVA) shows the term to be significant at $p < 0.0001$—a less than 1:10,000 chance of it being this large if the true effect were null. (For a primer on ANOVA and p-values, refer to *DOE Simplified*.) So together gold and copper melt at a lower temperature than either one alone—isn't that amazing!

OTHER DEPRESSING COMBINATIONS OF MATERIALS

Depressed m.p. from mixtures of one material with another, such as gold with copper, are not that uncommon. The point at which a mixture of two such substances reaches the minimum melting temperature is called the "eutectic." For example, an ideal solder for electronic circuitry is made from 63% tin (m.p. 450°F) and 37% lead (m.p. 621°F)—together these metals melt at a lower temperature (361°F) than either one in pure form. The constituents crystallize simultaneously at this temperature from molten liquid solution by what chemists call a eutectic reaction. The term comes from the Greek eutektos, meaning "easily melted."

The most prevalent eutectic reaction that we encounter in Minnesota occurs when our highway workers spread salt on roads to aid snow removal. The eutectic point for sodium chloride occurs at 23.3 wt% in water at a freezing point of minus six degrees Fahrenheit (−6°F). As salt is added to a mixture of water and ice on winter roads, some of the ice melts due to the depression of the m.p. That causes heat to be absorbed from the asphalt or concrete surface, which is no big deal—it's got lots to give. However, in a well-insulated environment like the jacket of an old-fashioned ice-cream maker this effect becomes very chilling (and useful!).

Mathematically, due to the coding on a zero to one scale for each component, the maximum impact of this second-order effect (x_1x_2) occurs at the 0.5–0.5 ("50/50") blend. Some quick figuring will help you see that this must be so. First, multiply 0 by 1 and 1 by 0 to get the products at either end of the scale. If you do not compute zero in both cases, then perhaps you possess the street smarts to be a vendor like the one we quote in the sidebar below. Now things get a lot harder because fractions are involved. Multiply ¼ by ¾ and ¾ by ¼ to work out the result for the two axial check blends that this design specifies the centroid and the vertices. If you got the first calculation, we trust you know that either way this product comes to three-sixteenths. This is a little less than the 1/4 of the result you get from multiplying 0.5 by 0.5 for the "50/50" blend at the centroid.

A FISHY WAY TO BLEND 50/50

A New Orleans street vendor was asked how he could sell Gulf shrimp-cakes so cheap. "Well," he explained, "I have to mix in some big old Mississippi catfish that the trawler dredges off the river bottom when it makes a shrimp run. Nevertheless, I mix them up 50:50—one shrimp, one catfish."

If you look closely at the curve in Figure 1.2, you may notice that the minimum actually occurs just a little to the right of the 0.5–0.5 point. This is due to the gold having a lower m.p. than the copper, thus favoring a bit more of this noble metal. A computerized search for the minimum using a hill-climbing algorithm finds the minimum at 0.55 weight fraction gold, and thus 0.45% copper is required to make the two components total to 1.

Now for a major disclaimer: A mixture experiment like this one on gold and copper will only produce an approximation of the true response surface—it may not be accurate, particularly for the fine points such as the eutectic temperature. In the end, you must ask yourself as a formulator whether the results can be useful for improving your recipe. In this case, the next step would be to select a composition that meets the needs of solder for goldsmithing fine jewelry. Determine the predicted m.p. from the graph or more precisely via the mathematical model. Then run a confirmation test to see how close you get. As a practical matter, this might be off by some degrees and yet still be useful for improving your process.

TRIAL OF THE Pyx

In Anglo Saxon times the debasing of gold coin was punished by the loss of the hand. In later years, the adulteration of precious metals was prohibited by the Goldsmiths' Company of London (founded 1180). The composition of gold sovereigns was fixed eventually at eleven-twelfths

(Continued)

fine gold, and one-twelfth alloy (copper). So accurate became the composition and weight of the coin issued from the mint that at the 1871 trial of the "Pyx" the jury reported that every piece they separately examined, representing many millions of pounds sterling, was found to be accurate for both weight and fineness. The term "Pyx," Greek in origin, refers to the wooden chest in which newly-minted coins are placed for presentation to the expert jury of assayers assembled once a year at the Hall of the Worshipful Company of Goldsmiths in the United Kingdom. This ceremony dates to 1282.

Source: Encyclopaedia Britannica, 10th Edition (1902).

Details on Modeling the Performance of a Two-Component Mixture

Our example of blending glittery metals for jewelry provides a specific application of mixture design and modeling. Now that we've enticed you this far, it's time to consider some general guidelines for setting up a formulation experiment and analyzing the results. Let's start with the Scheffé equations for predicting the response from two components.

$$\text{First order (linear): } \hat{y} = \beta_1 x_1 + \beta_2 x_2$$

$$\text{Second order (quadratic): } \hat{y} = \beta_1 x_1 + \beta_1 x_1 + \beta_{12} x_1 x_2$$

The hat (^), properly known as a circumflex, over the letter y symbolizes that we want to predict this response value. The β (beta) symbols represent coefficients, fitted via regression.

We detail the third order (cubic), which you may never need, in the Appendix 1A. There, for added measure, we also spell out the fourth-order (quartic) Scheffé equation. By this stage, very complex behavior can be modeled for all practical purposes. However, this process of model-building could continue to infinite orders of the inputs x to approximate any true surface in what mathematicians refer to as a Taylor polynomial series.

HOW TO ORDER UP JUST THE RIGHT MODEL FOR YOUR MIXTURE DATA

Refer to Appendix 1B for clever statistical procedures that sequentially test polynomial orders, layer by layer, for significance and lack of fit. This process produces a model that provides an adequate explanation of the results with as few predictor terms as possible, which is, "parsimonious." We spell it out for the gold jewelry case. Do not be put off by mathematically-intense statistical procedures—software handles all the calculations. You just need to get comfortable with the outputs and competent in concluding which model they favor.

> The mentally disturbed do not employ the Principle of Scientific Parsimony: the simplest theory to explain a given set of facts. They shoot for the baroque.
>
> **—Philip K. Dick ("Valis," p. 16, Mariner Books, 2011)**

The second-order equation not only may suffice for your needs to characterize the two primary components in your formulation, but it also could reveal a surprising nonlinear blending effect. The possibilities are illustrated graphically in Figure 1.3, which presumes that the higher the response, the better.

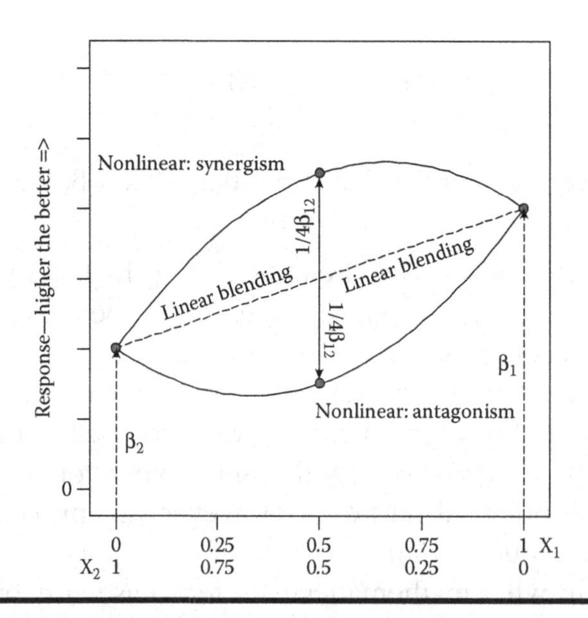

Figure 1.3 Graphical depiction of second-order mixture-model.

Notice that we tilted the linear blending line upwards, in other words, β_1 exceeds β_2. So this response surface predicts better performance for pure x_1 than for pure x_2. If together these two ingredients produce at the same rates as when working alone, then at the 0.5–0.5 midpoint the response will fall on the linear blending line. However, you hope that they really hit it off and produce more than either one alone. Then the response will curve upwards—producing the maximum deflection at the midpoint. This synergistic (positive) nonlinear blending effect equals one-fourth (0.5 * 0.5) of the second-order coefficient. Unfortunately, some components just do not work very well together, and things get antagonistic. Then the response curves downward and the β_{12} coefficient becomes negative.

ISOBOLOGRAMS

In 1871 T.R. Fraser introduced a graphical tool called the "isobologram." It characterized departures from additivity between combinations of drugs. Although it differs a bit in shape from our graph in Figure 1.2, the isobologram is essentially equivalent—it plots the dose-response surface associated with the combination superimposed on a plot of the same contour under the assumption of additivity, that is—linear blending. The observed results are called the "isobol," generally produced for the combinations of individual drug dosages that produce a 50% response by the subjects. If the isobol falls below the line of additivity, a synergism is claimed, because less of the drugs will be needed. On the other hand, if the isobol rises above the line, then the drugs are presumed to be antagonistic. However, there are two major shortcomings associated with the use of isobolograms. They do not account for data variability, and they are restricted to only a few components.

Source: Meadows, S.L. et al., *Environ. Health Perspect.*, 110, 979, 2002.

In this example, we made the response one where higher is better. Thus, a positive β_{12} coefficient is desirable for this nonlinear blending effect. However, in the first example—blending of copper into gold—the negative nonlinear coefficient was what the jewelry maker hoped to see. Thus, a synergistic deflection off the linear blending slope on the response surface could be positive or negative, depending on the goal being maximization or minimization.

When experimenting with mixtures, it pays to design an experiment that provides enough unique blends to fit the second-order Scheffé polynomial model. Then you can detect possible nonlinear blending effects. Hopefully, these will prove as advantageous as the synergistic combination of copper with gold for soldering fine jewelry. Meanwhile, it's just as well to know about antagonistic ingredients so you can keep these separated!

AN ANALOGY FOR NONLINEAR BLENDING THAT WILL WORK FOR YOUR MANAGER

In the go–go years of the computer industry of the late 1990s (before the dot-com bust) a movement developed for extreme programming (XP)—a form of agile development. One of its premises, which many software executives found counter-intuitive is that two people working at a single computer are just as productive as they would be if kept in separate cubicles, but (this is the payoff!) this pair programming increases software quality. Thus, better code emerges without impacting time to deliver. Wouldn't that be a wonderful case of synergistic (nonlinear) blending?

The idea of creating teams of two is nothing new. It's easy to imagine cavemen pairing up to hunt down mastodons more effectively. However, an age-old problem for managers of such tasks, for example, a police chief setting up patrol cars, is choosing the right people to team together. Unfortunately, some combinations turn out to be antagonistic, that is—they produce less as a partnership than either one alone. That creates several headaches all around.

How does one know which elements will prove not only to be compatible but also—create a synergy? Experiment!

Practice Problems

To practice using the statistical techniques you learned in Chapter 1, work through the following problems. Statistical software used for such computations can be accessed freely via a website developed in support of this book. There you will also find answers posted. See "About the Software" for instructions.

A LIFE PRESERVER FOR THOSE WHO BELIEVE STATISTICS IS OVER THEIR HEADS

Our first two books, *DOE Simplified* and *RSM Simplified*, offer all the statistical tools that one needs to choose predictive models and validate them statistically. However, leaving nothing to chance, we offer you readers a stimulating web-based primer on the fundamental statistics used in the design of experiments. This dynamic, interactive presentation, called "PreDOE" will fill in gaps on analysis of variance (ANOVA), p-values, and so on. It normally requires a fee, but Stat-Ease will waive this for readers of the Simplified series by Anderson and Whitcomb. If you are not sure whether you need to invest time in PreDOE, which may engage your attention for a number of hours, start with the online self-assessment on statistics for the design of experiments. To get the link and validation, email workshops@statease.com with reference to the page number of this offer for free access to PreDOE web-based training.

> Errors, like straws, upon the surface flow; He who would search for pearls must dive below.
>
> **—John Dryden (1678)**

> When p is low, null must go. When p is high, null will fly.
>
> **–Author unknown**

P-values are the calculated probability, used to evaluate statistical significance in a hypothesis test.

Problem 1.1

To reinforce the basics of mixture modeling presented in this chapter, we will start you off with some obvious questions that stem from this imaginary, but commonplace, situation in our heartland of the United States.

The old truck on your hobby farm gets very poor gas mileage. Luckily you can purchase fuel from a wholesaler who serves the agricultural market—A low-grade gasoline that produces 10 miles to the gallon (mpg) then it's alright to drive the old truck all the way back into the city where you usually dwell. It's cheap; only 3 dollars a gallon. Another possibility is to purchase the highly refined premium gasoline that increases the engine efficiency to 14 mpg. However, it costs 4 dollars a gallon.

Consider these questions:

1. Assuming you drive 1,000 miles per year going back and forth from your hobby farm, which grade of gasoline should you buy to minimize your annual fuel cost?
2. Now suppose the wholesaler offers to blend these two fuels 50/50 at $3.55 per gallon: How does this differ from the linear blend of prices?
3. Furthermore, you discover that your old truck gets 13 mpg with this blend of gasoline: Is this a synergism for fuel economy?
4. Should you buy the 50/50 blend of the two grades of gasoline? (Do not assume this will be so. Even if synergism is evident, the beneficial deflection off the linear blending point may not achieve the level of the best pure component. However, in this case, the solution requires an economic analysis—look for the best bottom line on costs per year.)

INVERSE TRANSFORMATION PUTS MILEAGE COMPARISONS ON TRACK

When the price of gas went over 4 dollars a gallon, I started paying attention to which of my three cars went where. For example, my wife and her sister traveled 100 miles the other day to do some work at the home of their elderly parents. They had our old minivan loaded up, but, after thinking about it getting only about 15 miles per gallon (mpg), I moved all the stuff over to my newer Mazda 6 Sports Wagon, which gets 25 mpg. That meant no zoom-zoom for me that day going to work, but it was worth enduring the looks of scorn from the other road warriors.

National Public Radio's (NPR) All Things Considered show on June 19, 2008, led off with this quiz: "Which saves more gas: trading in a 16-mile-a-gallon gas guzzler for a slightly more efficient car that gets 20 mpg? Alternatively, going from a gas-sipping sedan of 34-mpg to a hybrid that gets 50 mpg?" Of course, the counter-intuitive answer is the one that's correct—the first choice.

This is referred to as a "math illusion" studied by Richard Larrick, a management professor at Duke University. He found it easy to fool college students into making the wrong choice in puzzlers like that posed by NPR. Larrick suggests that it makes far more sense to report fuel efficiency in terms of gallons per 10,000 miles (gpm)—an average distance driven per year by

(*Continued*)

the typical USA car owner. Professor Larrick was inspired to promote "gpm" (vs. mpg) after realizing in the end that he'd be better off trading in the family minivan and only gaining 10 miles per gallon with a station wagon, rather than swapping his second car, a small sedan, for a highly efficient hybrid.

Are you still not sure about the NPR puzzler? Imagine you and your spouse work at separate locations that require an annual commute of precisely 10,000 miles per year for both of you driving separately (two automobiles). Then your 16-mpg guzzler consumes 625 gallons (10,000/16). By trading that for a 20-mpg car, you will need only 500 gallons the next year—a savings of 125 gallons. On the other hand, your spouse drives the far more efficient 34 mpg sedan—it requires only 294 gallons of gas per year (10,000/34). Upgrading this to the 50-mpg hybrid saves just 94 gallons! We will let you do the math on this last bit.

It is surprising how something as simple as an inverse transformation makes things so much clearer.

PS: For more details on transformations, refer to Chapter 4, "Dealing with Nonnormality via Response Transformations" in the 3rd edition of *DOE Simplified.*

Problem 1.2

This exercise stems from an experiment done by Mark with help from his daughter Katie. To demonstrate an experiment on mixtures, they blew up a plastic film canister—not just once, but over a dozen times. The explosive power came from Alka Seltzer®—an amalgam of citric acid, sodium bicarbonate (baking soda) and aspirin (Figure 1.4).

You can see the experimental apparatus pictured: launching tube, a container with water, the tablets, plastic film canister (Fuji's works best), a scale and stop-watch. Research via the Internet produced many write-ups on making Alka Seltzer "rockets." These are generally recommended when using only a quarter of one tablet, and they advocate experimentation on the amount of water, starting by filling the canister halfway. Mark quickly discovered that the tablets break apart very easily, so he found it most convenient and least variable to simply put in a whole tablet every time (a constant). It then took a steady hand to quickly snap on the top of the canister, over which Katie placed the launching tube and Mark prepared to press his stopwatch. (Subsequent research on this experiment indicated it would have

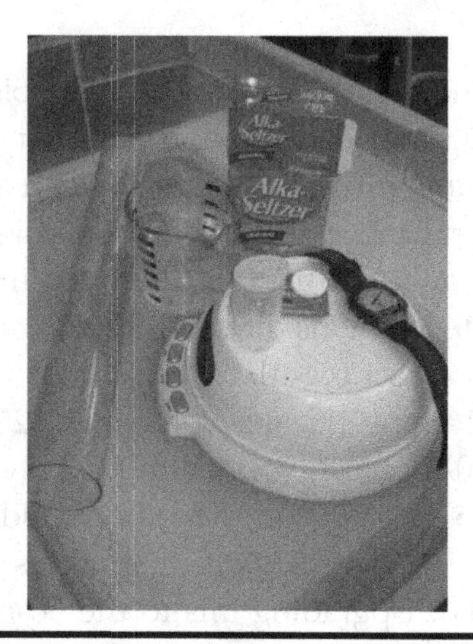

Figure 1.4 Apparatus for film-canister rocketry.

been far less nerve-wracking to stick the tablet on the lid with chewing gum, put water in the container, put the lid on, and then tip it over—shooting the canister into the air.) After some seconds the explosion occurred—propelling the lid from the back porch to nearly the roof of his two-story home.

HEADS UP—DO NOT PICK PRANKSTERS AS YOUR ASSISTANT ON ROCKET SCIENCE

Those of you who are fans of Gary Larsen's Far Side series of cartoons may recall a classic on depicting a white-coated scientist putting the last nail on the nosecone of a big rocket. In the background you see his assistant sneaking up with an inflated paper bag—poised to pop it! Mark's rocketry assistant, Katie, discovered that enough fizz remained in the canister to precipitate the second blow up. On randomly chosen runs she would sneak up on her father while he recorded the first shot's results and blast away. The only saving grace for Mark was the ready availability of Alka Seltzer for an ensuing headache.

Before designing this experiment, Mark did some range finding to discover that only 4 cubic centimeters (cm^3) of water in the 34 cm^3 canister would produce a very satisfactory explosion. However, it would not do to fill the container completely because the Alka Seltzer effervesced too quickly and prevented placement of the lid. After some further fiddling, Mark found that a reasonable maximum of water would be 20 cm^3—more than half full.

He then set up a two-component mixture design that provided the extreme vertices (4–20 cm^3 of water), the centroid (12 cm^3) and axial check blends at 8 and 16 cm^3. Mark replicated the vertices and centroid to provide measures of pure error for testing lack of fit.

Just for fun, Mark asked several masters-level engineers, albeit not rocket scientists, but plenty smart, what they predicted—the majority guessed it would make no difference how much water given a minimum to wet the tablet and not so full it would prevent the top going on. This becomes the null hypothesis for statistical testing—assuming no effect due to changing the mix of air and water in the film canister.

Table 1.2 shows the results of flight time in seconds for various blends of water versus air. Looking at the data, sorted by the amount of water, do you agree with these engineers that this component makes no difference? You may be somewhat uncertain with only an "intraocular test"—statistical analysis would be far more definitive to assess the significance of the spread in flight times relative to the variation due to blending errors and the perilous process of launching the rockets. Now would be a good time to fire up your favorite statistical software, assuming it provides the capability for mixture design, modeling, analysis, response surface graphics, and multiple-response optimization. In case you have no such program readily available, we offer one via the Internet—see the "About the Software" section for the website location and instructions for downloading. To get started with the software, try reproducing the outputs embedded in the answer to this problem posted at the same site (in portable document format—.pdf). In the following chapters, we will lead you to more detailed tutorials on using this particular DOE program.

Table 1.2 Results from Film-Canister Rocket Experiment

Blend #	*Run*	*Type*	*A: Water (cm^3)*	*B: Air (cm^3)*	*Flight Time (s)*
1	2	Vertex	4	30	1.88
2	6	Vertex	4	30	1.87
3	4	AxialCB	8	26	1.75
4	3	Center	12	22	1.60
5	8	Center	12	22	1.72
6	5	AxialCB	16	18	1.75
7	1	Vertex	20	14	1.47
8	7	Vertex	20	14	1.53

BLASTING OFF FROM TUCSON, ARIZONA

After touring the Titan Missile Museum south of Tucson, Arizona, Mark found the toy pictured in their gift shop. This product, made by a local inventor (CSC Toys LLC), improves the aerodynamics of the seltzer-powered rocket by the addition of a nose cone and fins (Figure 1.5).

Like these film canister rockets, the thrust of the Titan missile depended on two components, albeit many orders of magnitude more powerful—a precisely controlled combination of nitrogen tetroxide (oxidizer) and hydrazine (fuel) that spontaneously ignited upon contact. This extreme exothermic chemical behavior is characterized as "hypergolic." The fuels were stable only at 58°F–62°F, which meant that temperature control was critical. In 1980 a worker dropped a 9-pound socket from his wrench down a silo and punctured the fuel tank. Fortunately, the 8,000-pound nuclear warhead, more destructive than all the bombs exploded in all of World War II, landed harmlessly several hundred feet away. Some years later the Titans were replaced with MX "Peacekeeper" rockets that used solid fuel.

Figure 1.5 The MIGHTY Seltzer Rocket pictured from a launch pad in Tucson.

Appendix 1A: Cubic Equations for Mixture Modeling (and Beyond)

The full cubic (third order) equation for modeling a two-component mixture is shown below:

$$\hat{y} = \beta_1 x_1 + \beta_2 x_2 + \beta_{12} x_1 x_2 + \delta_{12} x_1 x_2 (x_1 - x_2)$$

Notice that the coefficient on the highest order, non-linear blending term is distinguished by the Greek letter delta. Think of the letter "d" (delta) as a symbol for the differences ("d" for difference) pictured in Figure 1A.1. It depicts a very unusual response surface for two components with only first and third order behavior—the second-order coefficient was zeroed out to provide a clearer view of how the new term superimposes a wave around the linear blending line. Also, to add another wrinkle (pun intended) to this surface, the coefficient is negative.

$$\hat{y} = \beta_1 x_1 + \beta_2 x_2 - \delta_{12} x_1 x_2 (x_1 - x_2)$$

See if you can bend your brain around this complex mixture model: It's challenging!

At the 50–50 blend point the components are equal, so the offset is zeroed ($x_1 - x_2 = 0$). When x_1 exceeds x_2 to the right of the midpoint,

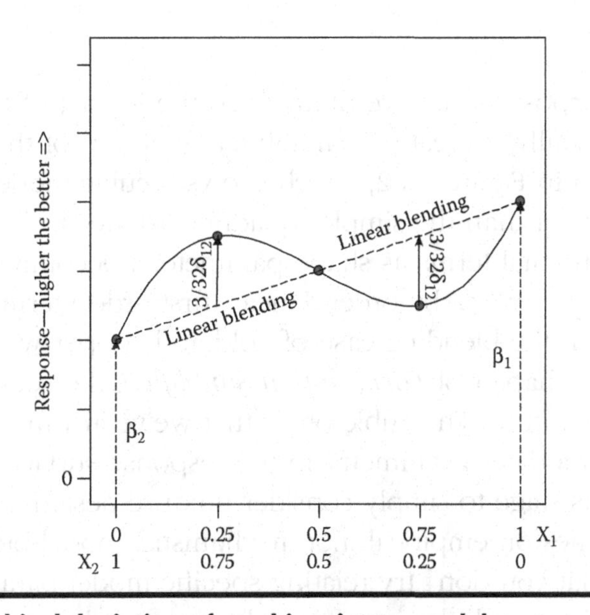

Figure 1A.1 Graphical depiction of a cubic mixture model.

the difference δ (delta) is positive; thus (due to the negative coefficient) the curve deflects downward. To the left, the difference δ becomes negative causing the response to go up (a negative times a negative makes a positive!). The maximum wave height from the linear blending of 3/32 delta occurs at one-fourth (1/4) and three-fourths (3/4) blend points (¾ * ¼ * (¾ – ¼) = 3/32).

FOR THOSE OF YOU MORE FAMILIAR WITH RSM MODELS

You may wonder why the third-order equation for mixtures is more complex than that used to model similar behavior in a process. Remember that the coding for process models goes from –1 to +1. When you cube these quantities, the positives stay positive (+1 * +1 * +1 = +1), and the negative stays negative (–1 * –1 * –1 = –1). Thus, you get wavy, up-and-down, behavior in the surface. But in mixtures, the coded units are 0 to 1, which will always be positive, so you model waviness via a difference of components.

PS: Since this chapter has kept things simple by focusing only on two components, the cubic mixture model is missing one general term that involves three components: $x_ix_jx_k$. You will see this term highlighted in future chapters that delve into a "special" cubic model, which shortcuts some unnecessary complexities and thus makes things a lot easier for formulators.

For illustrative purposes, only, we dramatized the impact of the cubic term in Figure 1A.1. Usually, it creates a far subtler "shaping" of the surface such you see illustrated in Figure 1A.2, which shows a cubic model (solid line) fitting noticeably better than the simpler quadratic (dotted).

Think of polynomial terms as shape parameters, becoming subtler in their effect as they increase by order. Linear (first order) terms define the slope. As shown in the blending case of gold and copper we went through earlier in this first chapter of *Formulation Simplified*, the second order (quadratic) fits curvature. The cubic order that we've just introduced in the Appendix accommodates asymmetry in the response surface.

It's good at this stage to simply consider mixture design as a special form of RSM, which relies on empirical, not mechanistic, model building. In other words, it's best that you don't try relating specific model parameters to the underlying chemistry and physics of your formulation behavior. However,

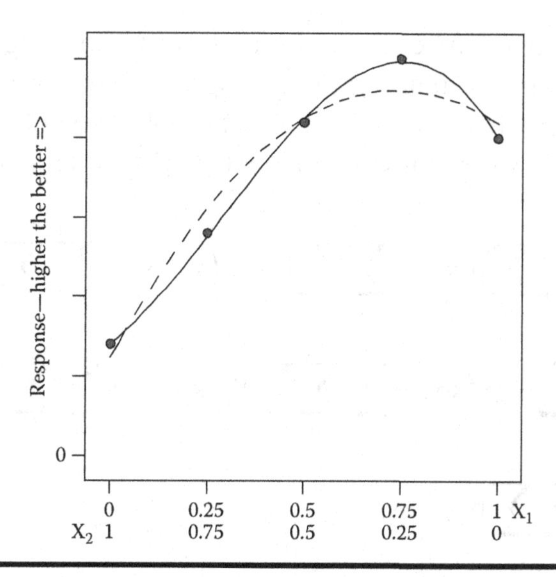

Figure 1A.2 A subtler surface fitted best with a cubic model.

asymmetry is more prevalent in mixture experiments than it is in process experiments. Thus, if the data suffices to fit a surface very precisely to the third-order model, it might capture subtle non-linearity that the quadratic would not; that is, this lower equation would exhibit a significant lack of fit.

AN IDIOM FOR NONLINEAR BLENDING: "CHEMISTRY"

Chemistry is the study of matter and its interactions. It amazes us by unexpected reactions between particular substances. The word "chemistry" is often used to expressively describe a potently positive pairing, such as an irresistible attraction between two lovers. Less often one may hear of a "bad chemistry" building up in a group that includes antagonistic elements. Thus this term "chemistry" has become a word that generally describes non-linear blending effects.

"We have very good chemistry team this year". Phil Housley, the Minnesota hockey great, assessing his 2009 Stillwater Area (Minnesota) High School team.

In any case, to fit this cubic equation, one must design an experiment with at least four unique blends, whereas three suffices (at the bare minimum) to fit the quadratic. The more complex the behavior you want to model, the more work you must do as a formulator. You get what you pay for.

If you have plenty of materials and time to mix them together—not to mention the capability for making many response measurements, you could design an experiment that fits a quartic Scheffé polynomial model. Here it is in general terms for however many components ("q") you care to experiment on:

$$\hat{y} = \sum_{i=1}^{q}\beta_i x_i + \sum_{i<j}^{q-1}\sum_{j}^{q}\beta_{ij}x_i x_j + \sum_{i<j}^{q-1}\sum_{j}^{q}\delta_{ij}x_i x_j\left(x_i - x_j\right) + \sum_{i<j}^{q-1}\sum_{j}^{q}\gamma_{ij}x_i x_j\left(x_i - x_j\right)^2$$
$$+ \sum_{i<j}^{q-2}\sum_{j<k}^{q-1}\sum_{k}^{q}\beta_{iijk}x_i^2 x_j x_k + \sum_{i<j}^{q-2}\sum_{j<k}^{q-1}\sum_{k}^{q}\beta_{ijjk}x_i x_j^2 x_k + \sum_{i<j}^{q-2}\sum_{j<k}^{q-1}\sum_{k}^{q}\beta_{ijkk}x_i x_j x_k^2$$
$$+ \sum_{i<j}^{q-3}\sum_{j<k}^{q-2}\sum_{k<l}^{q-1}\sum_{l}^{q}\beta_{ijkl}x_i x_j x_k x_l$$

Notice that squared terms now appear. Although statistical software (such as the one we provide to you readers) will handle the design and analysis of a mixture experiment geared to this fourth order, it is very unlikely that this will provide any practical gain over the fit you get from cubic or quadratic models. For response surface modeling, it's good to keep in mind the principle of parsimony, which advises that when confronted with many equally accurate explanations of a scientific phenomenon it's best to choose the simplest one (Anderson and Whitcomb, 2005, Chapter 1, sidebar "How Statisticians Keep Things Simple").

OUT OF ORDER?

Back in the days when computer-aided mixture modeling was limited to cubic, an industrial statistician cornered Mark at a conference and complained that he needed quartic to fit a formulation over the entire experimental region. Quadratic fit fine for most of the results but fell short where the performance fell off very rapidly. Mark tried a trick that his doctor told him after he injured his shoulder playing softball. "When does it hurt," the medico asked. "Only when I throw a softball," said Mark. "Just don't do that," the doctor advised. In similar fashion, Mark—being ever practical—suggested that one could simply not look at the response surface where it drops off and gets fit inaccurately because no one cares at that point.

(*Continued*)

This flippant advice is more helpful than you might think. If you can apply your subject matter knowledge and do some pre-experimentation to restrict the focus of the mixture design to a desired region, the degree of Scheffé polynomial required to approximate the response surface will likely be less, thus reducing the number of blends required by simplifying the modeling needed for adequate prediction power. For example, why model all the Rocky Mountains when you are really interested only in exploring one of the peaks?

Some might say that this question is academic, but that's OK because I am an academic.

—Kevin Dorfman, a Professor at the University of Minnesota, speaking on esoteric research in his specialty—chemical engineering on the macromolecular scale

Appendix 1B: Statistical Details on How to Order Up Just the Right Model

As noted in the sidebar "How to Order Up Just the Right Model for Your Mixture Data," it's best that predictive models be kept as parsimonious as possible. We refer to this as the KISS Principle, that is, Keep It Simple, Statistically. It can be accomplished via a statistical table called "sequential model sum of squares" (SMSS). Refer to Chapter 4 (pp. 76–77 and 96–98) of *RSM Simplified* for details on this process for fitting polynomial process models, which works equally well for mixture data. For example, Table 1A.1 shows the SMSS for the gold–copper blending case.

The SMSS provides an accounting of variation and associated p-values (Prob > F). To KISS, you should add a higher-level source of terms (i.e., order) only if it explains a significant amount of variation (i.e., $p < 0.05$) beyond what's already removed by lower orders. In this case, it appears at first that the model may not rise above the mean, there being no significance to the linear terms due to the p of 0.385 being far above the 0.05 threshold. However, the next level of quadratic comes out highly significant, so much so that the exact p-value is not worth noting, it being less than 1 in 10,000 (0.0001) probability that the 498 F value could occur by chance.

Table 1A.1 Sequential Model Sums of Squares

Source	*Sum of Squares*	*Df*	*Mean Square*	*F Value*	*p-Value Prob > F*
Mean versus total	$1.003 * 10^7$	1	$1.003 * 10^7$		
Linear versus mean	3542	1	3542	0.84	0.3850
Quadratic versus linear	33094	1	33094	498.0	<0.0001
Cubic versus quadratic	1.996	1	1.996	0.026	0.8775
Quartic versus cubic	30.71	1	30.71	0.36	0.5772
Residual	432.4	5	86.48		
Total	$1.007 * 10^7$	10	$1.007 * 10^6$		

Furthermore, because the quadratic term is significant, all the linear terms come back into the model to maintain hierarchy, which we explain in Chapter 5 of *DOE Simplified* in our sidebar (p. 103) on "Preserving Family Unity." To put it simply, parents must always be included with their children. Therefore, in this case, where the quadratic term AB merits inclusion in the model, both A and B must come back in for support, even though these linear terms, on their own are insignificant. It would be especially problematic in the case of mixture not to include the main ingredients in the model. That would not make any sense.

The computational details on the SMSS displayed in Table 1A.1, are shown diagrammatically in Figure 1A.3.

Testing for Lack of Fit

The lack-of-fit (LOF) test compares the deviation of actual points from the fitted surface, relative to pure error. If a model has a significant lack of fit, it should be investigated before being used for prediction (refer to Chapter 3, pp. 64–65, of *RSM Simplified* for details). You can see in Figure 1A.4 that the linear model does not fit nearly as well as the quadratic for the melting point of gold–copper blends.

The deviations of the points from the fitted line, indicated by the arrowed lines, obviously far exceed what one would expect from the variations within the replicated points for pure gold (left) and pure copper (right).

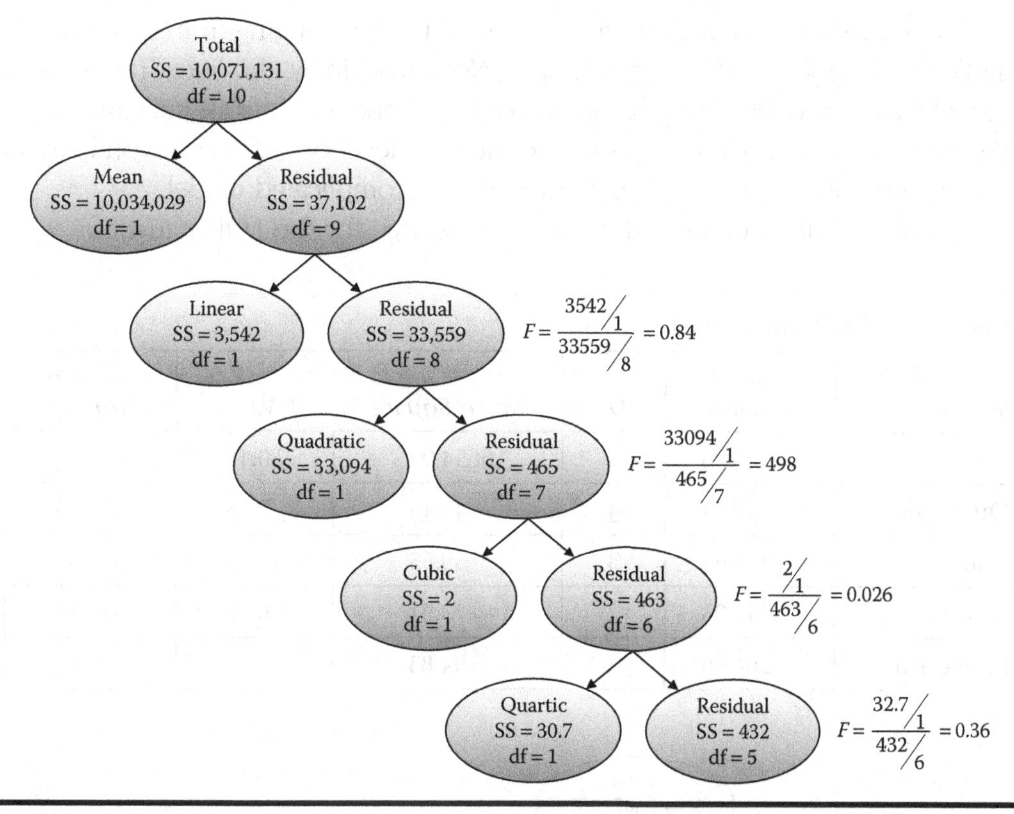

Figure 1A.3 Calculations for sequential model sums of squares in Table 1A.1.

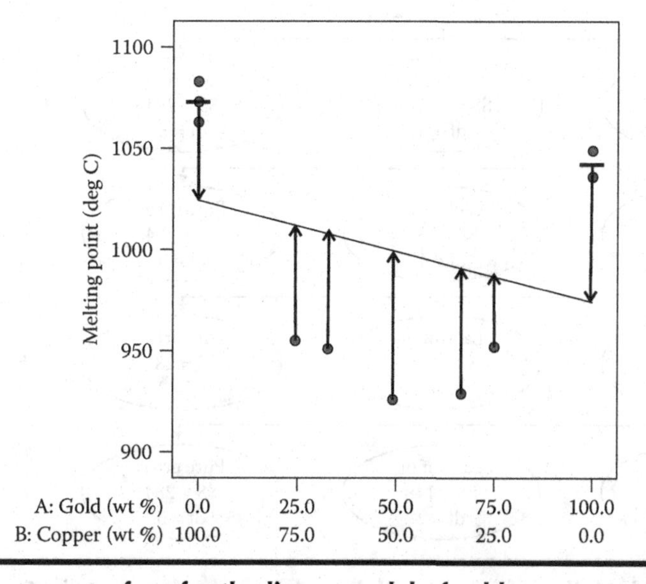

Figure 1A.4 Response surface for the linear model of gold–copper melting points.

Table 1A.2 shows the lack of fit statistics for models ranging from linear to quartic for the gold-copper blending case. Not surprisingly, the linear model gets rejected by the p-value being far below the 0.05 benchmark for significance. The quadratic model wins out. Going to the next level of cubic or beyond, an order which does no good—thus creates an overcomplicated model.

Figure 1A.5 diagrams the derivation by source for the LOF statistics.

Table 1A.2 Lack-of-Fit Tests

Source	*Sum of Squares*	*Df*	*Mean Square*	*F Value*	*p-Value Prob > F*
Linear	33274.76	5	6654.95	70.18	0.0026
Quadratic	180.64	4	45.16	0.48	0.7570
Cubic	178.64	3	59.55	0.63	0.6442
Quartic	147.93	2	73.97	0.78	0.5336
Pure error	284.50	3	94.83		

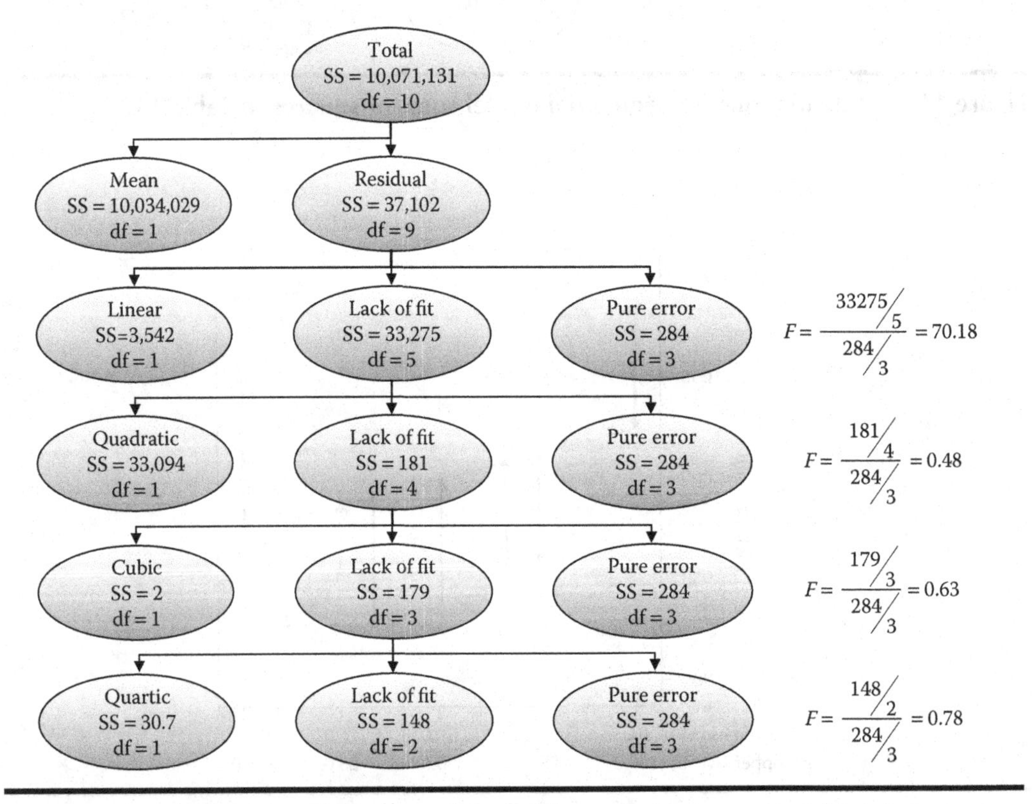

Figure 1A.5 Calculations for sequential model sums of squares in Table 1A.2.

Model Summary Statistics

Last, but not least, we turn our attention to the R-squared (R^2) statistic—a measure of fit that is very familiar to most experimenters. Table 1A.3 presents R-squared for the gold–copper blending experiment: as-is (raw), adjusted and predicted. As you can see by reading across row by row and noting the decreasing values, this series of calculations becomes more and more discriminating for model selection.

Table 1A.3 Model Summary Statistics

Source	*R-Squared*	*Adjusted R-Squared*	*Predicted R-Squared*
Linear	0.0955	−0.0176	−0.4734
Quadratic	0.9875	0.9839	0.9729
Cubic	0.9875	0.9813	0.9641
Quartic	0.9883	0.9790	0.9508

WHY R-SQUARED NEEDS TO BE ADJUSTED

We recommend you use either the adjusted or predicted R-squared—the raw value being flawed due to the fact that it leads to overfitting. The problem with R-squared is that it always increases. For example, going down the column of R-squared values in Table 1A.3 leads to the conclusion that the quartic model fits best. However, the adjusted R-squared increases only when you add good predictors. It decreases when you add useless terms to your model. For example, notice in Table 1A.3 how it goes down after adding the cubic term [AB(A–B)], and then it falls further with next order beyond that—the quartic term [$AB(A-B)^2$]. Therefore, there is no advantage of putting these complicated terms into the model for predicting melting point as a function of the relative amounts of copper and gold in the jeweler's blend.

Another difference of adjusted R-squared versus the raw statistic is that it can go negative, for example, as it does in this case for the linear model. That's not a good sign!

The predicted R-squared provides all the advantages of the adjusted, plus it serves better for experimenters who will be using their model to

(*Continued*)

produce better results in the future. This is accomplished by systematically removing each result, re-fitting the model, and measuring how far its prediction falls from the removed value.

The adjusted and predicted R-squared statistics usually lead to the same recommended model. However, when models contain many terms within a given order that are insignificant, the predicted R-squared can fall off so much that it leads to under-fitting. In that case, go with what the adjusted R-squared suggests, but try a model-reduction algorithm such as a backward stepwise elimination. Later in this book when we introduce combined mixture-process models we will consider this situation of discrepant adjusted versus predicted R-squared.

The quadratic model [A, B, and AB] comes out on top overall with the summary measures of the adjusted and predicted R-squared (never mind the raw R-squared!).

Chapter 2

Triangulating Your Region of Formulation

> If you don't know where you are going, you will wind up somewhere else.
>
> **—Yogi Berra**

In this chapter, we will build up from the simplicity of dealing only with two components system and then experiment on three or more. The biggest step will be recognizing that if you lay this out in rectangular coordinates then you really do not know where you are going and you will wind up somewhere else (to paraphrase baseball guru Yogi). You need to get yourself into the triangular space depicted in Figure 2.1.

The levels of three ingredients can be represented on this two-dimensional graph paper, also known as "trilinear" for the way it's ruled. It depicts blends of up to three materials:

1. Vertices are the pure components. For example, pure X_1 (or ingredient "A") is the point plotted at the top. For the sake of formulators, this paper is marked off on a zero to one-hundred scale which can be easily translated to a more mathematically convenient range of zero to one.
2. Sides are binary blends. Being a yardstick on two components, the sides are also referred to as "q-2 flats" (Myers et al., 2009, p. 570). The midpoints of these q-2 flats are 50/50 blends of the components at each end of the side. For example, the point between A and C represents exactly half of each (and none of material B!).

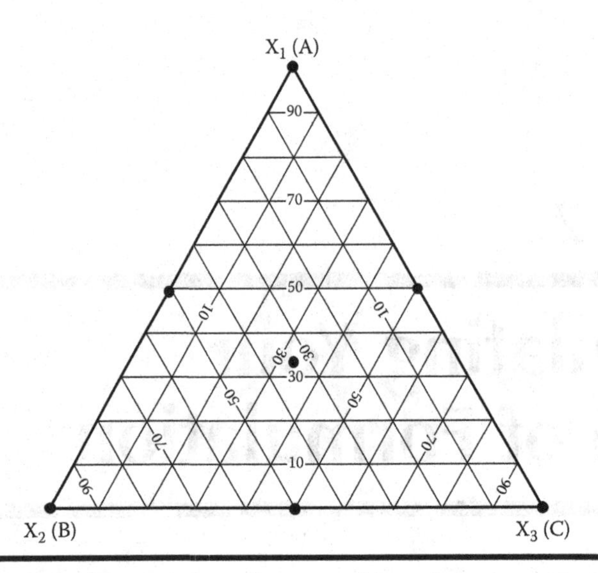

Figure 2.1 Trilinear graph paper for mixtures with points to mark pure components, binary blends, and overall centroid.

3. Mixtures of three components are in the center area. For example, the point located precisely in the middle of the triangle, called the "centroid," represents a blend of one-third each of all three ingredients.

THE INVENTION OF TRILINEAR PAPER

Michael Friendly and Daniel Denis in their "Milestones in the history of thematic cartography, statistical graphics, and data visualization" (2001) web document (www.datavis.ca/milestones/) attribute the invention of trilinear coordinates (graphs of (x, y, z) where x + y + z = constant) to the American thermodynamicist Josiah Gibbs in 1873, who used them for phase diagrams. They remain an enigma for many scientists who really ought to make use of trilinear paper.

> Triphase diagrams seem specially designed to confuse those who try to interpret them.
>
> **—Steven Abbott, "Practical Surfactants," "Phase Diagram Explorer," www.stevenabbott.co.uk/practical-surfactants/pde.php**

The neat thing about mapping mixtures to this triangular space is that once you know two component fractions, the third is determined by the total.

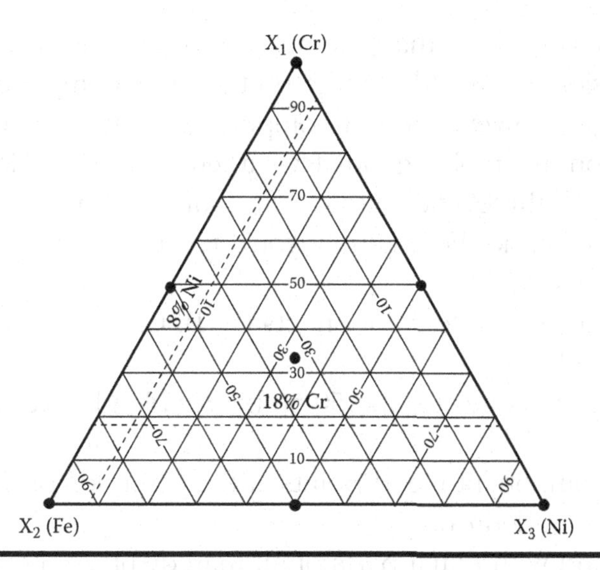

Figure 2.2 Locating the 18-8 composition of stainless steel (for flatware).

To illustrate how this works, consider the stainless-steel flatware—knives, forks, spoons, and so on—that you keep handy in your kitchen drawer for everyday eating. A very common metallurgical formulation for this purpose is 18% chromium (Cr) and 8% nickel (Ni) by weight—the remainder being iron (Fe), of course. Let's plot this on the trilinear paper laid out in Figure 2.2.

In this case, the metallurgist conveniently chose chromium as the first component to make the starting point easy—simply draw a line horizontally 18% of the way from the bottom (0% Cr) to the top (100% Cr). Now things get tricky because you must rotate the graph, so the nickel (Ni) comes out on top. (Think of these triangular graphs as being "turnery" paper!) Then it will be easy to draw the 8% line for this metal, designated as the third component. Now turn the graph, so iron (Fe) is at the top. Notice that the two lines intersect 74% of the way up from the zero base of iron to its pure component apex. The three ingredients now add to 100%! This feature of the ternary graph is very convenient for formulators.

The Simplex Centroid Design

The pattern of points depicted in Figure 2.1 forms a textbook design called a "simplex-centroid" (Scheffé, 1963; Cornell, 2002). We will introduce a more sophisticated design variation called a "simplex-lattice" later, but let's not get ahead of ourselves. The term "simplex" relates to the geometry—the simplest figure with one more vertex than the number of dimensions. In this case, only two dimensions are needed to graph the three components on to an equilateral triangle. However, a four-component mixture experiment requires another dimension in simplex geometry—a tetrahedron, which looks like a pyramid, but with three sides rather than four. To show how easy it is to create a simplex centroid, here is how you'd lay it out for four components:

1. Four points for the pure components (A, B, C, D) plotted at the corners of the tetrahedron).
2. Six points at the edges for the 50/50 binary blends (AB, AC, AD, BC, BD, CD).
3. Four three-component blend points at the centroids of the triangular faces of the tetrahedron.
4. The one blend with equal parts of all ingredients at the overall centroid of the tetrahedron.

This totals to 15 unique compositions from the four components. See these depicted in Figure 2.3.

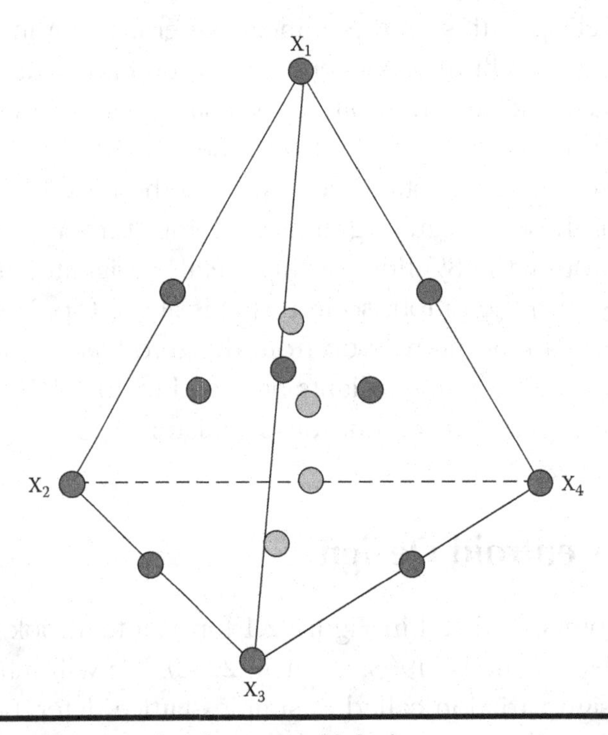

Figure 2.3 Four-component simplex centroid design.

Although the simplex-centroid is not a very sophisticated design, it does mix things up very well by always providing a blend of all the components (the overall centroid). Let's work through an example of a simplex centroid for three components, illustrating an entire cycle of mixture experimentation from design, through actual execution of the runs, statistical analysis, and finally, optimization of multiple responses with cost taken into account.

The Black and Blue Moon Beer Cocktail

A "cocktail" generally refers to a mixture of hard (high alcohol) liquor. Few drinkers nowadays would think of blending beers. However, one summer, Mark was hit by a sudden intuition that it might be very tasty to combine a black lager with a wheat beer—possibly this might produce a synergistic sensation. Furthermore, to provide a contrast to this premium pairing and possibly enable cost savings, he decided to mix in a cheap lager.

Once this idea took hold, Mark knew it must be tested by unbiased tasters with a talent for drinking beer, and that the experiment itself had to be conducted in a way that would prevent preconceived notions from contaminating the results. Let's see what can be learned via this case study about the application of multicomponent mixture design aimed at discovering a sweet spot of taste versus cost.

DO NOT ASSUME A DIRECT CORRELATION OF COST WITH QUALITY

In the late 1970s Mark took an evening course in marketing en route to his Masters in Business Administration (MBA). He worked through a case study showing how, although the national beer brands in the USA differed very little in their brews, their marketing campaigns divided drinkers into distinct segments. For example, Miller advertised their high-priced product as the "champagne" of bottled beer while Old Milwaukee went for the working man and took the low road on price. Meanwhile, on Mark's day job as a chemical process development engineer, an R&D colleague made a big deal over how one got what one paid for in a beer: The cheap stuff was simply swilled in his opinion. At this time, Mark was gaining a great appreciation for experiments based on statistical principles, such as the use of the null hypothesis for reducing prejudice.

(*Continued*)

Here was an opportunity to put the beer snob to the test via a blind, randomized, statistically-planned experiment. You can guess the outcome: He rated the Old "Swill"Waukee (his misnomer) number 1!

> It will come to pass that every braggart shall be found an ass.
>
> **—William Shakespeare**
> **(from All's Well that Ends Well)**

> He was a wise man who invented beer.
>
> **—Plato**

Here are the beer-cocktail ingredients (prices per 12 ounces, serving shown in parentheses):

1. Coor's brand Blue Moon Belgian-style wheat ale ($1.16)
2. Anheuser–Busch brand Budweiser American lager ($0.84)
3. Samuel Adams brand Black Lager ($1.24)

The design of the experiment, based on a simplex centroid, is laid out in Figure 2.4.

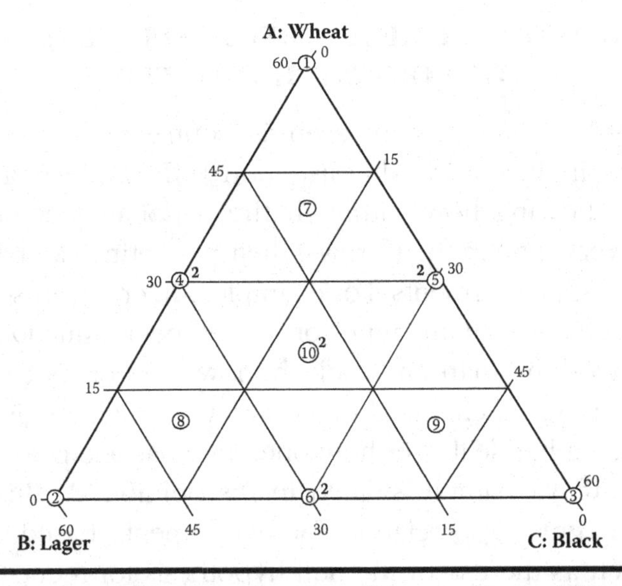

Figure 2.4 Simplex centroid bolstered with replicates and check blends.

Mark bolstered this design in three ways:

- Three added blends midway between the centroid and each of the vertices (pure beers). In the jargon of mixture design, these are called "axial check blends." They otherwise fill empty spaces in the experimental region. The addition of points to a textbook layout like the simplex centroid is called "design augmentation."
- Four point replicates (designated by "2"s)—the three binary blends (midpoints of sides) plus the centroid. This provides four measures ("degrees of freedom" in statistical lingo) of pure error. By establishing a benchmark against which the deviations of actual points from the fitted line can be assessed, pure error enables the testing of lack of fit—useful for assessing model adequacy.
- Three replications of the entire design were sampled by three tasters. Although the three subjects were chosen carefully based on their good taste in beer, they differed in their generosity of rating; that is, tending to score every brew higher or lower. These individual biases were corrected via a statistical technique called "blocking."

The 14 blends per person (blocked) were provided in random run order for these three sensory responses:

Y1. Taste (without looking at the cocktail) on hedonic scale of 1 (worst) to 9 (best)
Y2. Appearance (1–9)
Y3. Overall liking (1–9)

To keep things simple for educational purposes, we will only look at the overall liking (Y3) and the response of cost, which is determined completely by the blend's composition and, of course, the current cost of each ingredient.

Mark owns a very accurate kitchen scale (pictured in Figure 2.5) that he uses to weigh out green coffee beans for roasting (another story!) so it was convenient for him to set the total for each blend by weight rather than volume—to 60 grams (roughly two fluid ounces). That kept the total beer consumption per person to a reasonable level—about two bottles worth. (Mark admits that during the experiments he managed to drink about the same amount—in the name of science, naturally).

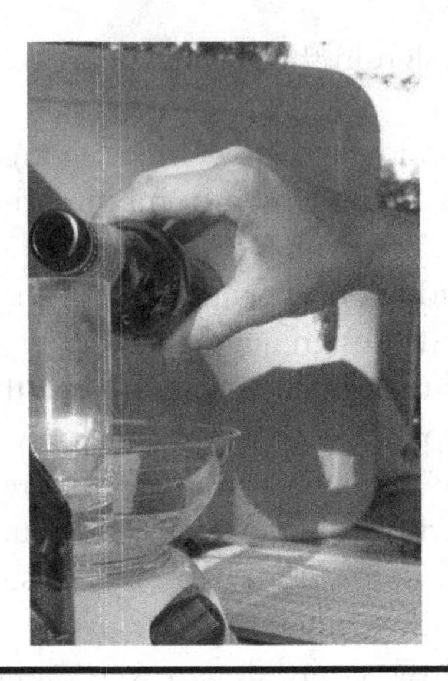

Figure 2.5 Precisely mixing a beer cocktail behind the screen (to keep tasters blind).

Each drinker kept his own beer-shot glass. The plan was for excess material, beyond what was required for measurement purposes, to be discarded, but Mark found this very difficult to enforce on the hot summer day of the experiment—done on his back porch.

EARLY THREE-COMPONENT BEER MIXTURES

According to popular bar lore, eighteenth-century Londoners developed a liking for a beverage called "three threads"—made by blending a third of a pint each of ale, beer and "twopenny" (the strongest of these brews, costing two cents a quart). Another three-component recipe dictated stale (aged for up to two years), mild and country (pale) ale. It seems likely that pub keepers experimented on mixtures in the hopes of finding something both tasty and cheap (by diluting costly brews with less expensive ones).

> It is not entirely clear as to why a fashion for mixing beers arose…, other than a desire to match palate and pocket.
>
> **—Beer Before Porter**
> **www.london-porter.com**

(Continued)

PS: Many Americans say "cheers" to a binary blend of beers from the British Isles—thick, dark stout poured on a pale ale. This is commonly called a "Black and Tan." What makes this combination interesting is that with the right order of addition—first the ale and then the stout, and a steady hand on pouring the drink, exhibits a distinct layering of black on tan. Evidently, the stout is less dense than the ale, but the reason remains mysterious—a matter for more research, no doubt!

Fear not the beer cocktail.

—Stephen Beaumont
World of Beer

See Table 2.1 for the text matrix, laid out by blend type and location, and the overall liking ratings for the three tasters. The actual order of presentation was randomized, thus decoupling the cocktail type from possibly lurking variables such as degrading taste (related to admissions above), dehydration from exposure to the summertime elements, and so on.

Be careful about drawing too many conclusions and extrapolating these very far. However, like all experiments, this one may produce some useful findings. Let's see what can be made of it.

Go ahead and look over the results—as Yogi Berra said "you can see a lot just by looking." For example, is it possible that some combinations of beers might be perceived as being unexpectedly tasty? Or, perhaps, the opposite may be true: Putting individual taste of certain beers together may not be such a good idea. Keep in mind that this experiment represents only a sampling of possible reactions by these particular tasters, who may or may not represent a particular segment of the beer-drinking market. Does it appear as if any of the three tasters may have been tougher than the others (hint!)? If so, do not worry; so long as this individual remains consistent with the others in his or her relative rankings by blend, then this consistent bias can be easily (and appropriately) blocked out mathematically, thus eliminating this easily-anticipated source of variation (person-to-person).

Table 2.1 Results of Beer-Cocktail Experiment (Ratings by Three Tasters by Blend)

Blend #	*Type*	*Location (Coded)*	*A: Wheat (Grams)*	*B: Lager (Grams)*	*C: Black (Grams)*	*Liking 1–9*	*Cost $/12 (ounces)*
1	Pure	Vertex (1,0,0)	60	0	0	5,5,5	1.16
2	Pure	Vertex (0,1,0)	0	60	0	4,4,3	0.84
3	Pure	Vertex (0,0,1)	0	0	60	7,6,5	1.24
4a	Binary	Center Edge (0.5,0.5,0)	30	30	0	5,5,4	1.00
4b	"	"	30	30	0	5,4,5	"
5a	Binary	Center Edge (0.5,0.5,0)	30	0	30	8,7,6	1.20
5b	"	"	30	0	30	7,8,7	"
6a	Binary	Center Edge (0.5,0.5,0)	0	30	30	4,4,3	1.04
6b	"	"	0	30	30	4,4,2	"
7	Check	Two-third, one-sixth,one-sixth Axial ($0.6\overline{6}$, $0.1\overline{6}$, $0.1\overline{6}$)	40	10	10	6,7,5	1.12
8	Check	One-sixth,two-third,one-sixth Axial ($0.1\overline{6}$, $0.6\overline{6}$, $0.1\overline{6}$)	10	40	10	5,5,4	0.96
9	Check	One-sixth,one-sixth,two-third Axial ($0.1\overline{6}$, $0.1\overline{6}$, $0.6\overline{6}$)	10	10	40	7,7,6	1.16
10a	Ternary	Centroid	20	20	20	5,7,4	1.08
10b	"	"	20	20	20	6,6,4	"

Diving Under the Response Surface to Detail the Underlying Predictive Model

The results are very interesting. As you can see by the location of the peak region in the 3D response plot (Figure 2.6), a blend of Blue Moon wheat ale (A) and Sam Adams Black Lager (C) really hit the spot for overall liking! Taste and appearance ratings also favored this binary blend. The tasters all preferred this combination, which Mark deemed the "Black and Blue Moon" beer cocktail.

The fitted model in coded units (0–1) that produced this surface is:

$$\text{Overall Liking} = 4.92A + 3.68B + 6.13C + 2.01AB + 7.35AC - 4.65BC$$

Notice from the coefficients on the pure-component terms (conveniently labeled alphabetically—A, B, and C, rather than mathematically—X_1, X_2, X_3) that these beer aficionados liked the Sam Adams Black Lager (C) best and Budweiser (B) the least: $6.13C > 4.92A > 3.68B$. From the second-order non-linear blending terms one can see the synergism of beer A (wheat) with beer C (black) by the large positive coefficient: 7.35AC. Do not get too excited by the very high coefficient: Remember that this value must be multiplied by one-fourth to calculate the "kicker" for the binary blend. Nevertheless, this product provides nearly a two-point gain on the hedonic scale: $7.35 * 1/4 = 1.84$.

On the other hand, these tasters' buds were antagonized by combining the Budweiser with the Sam Adams Black Lager as evidenced by the

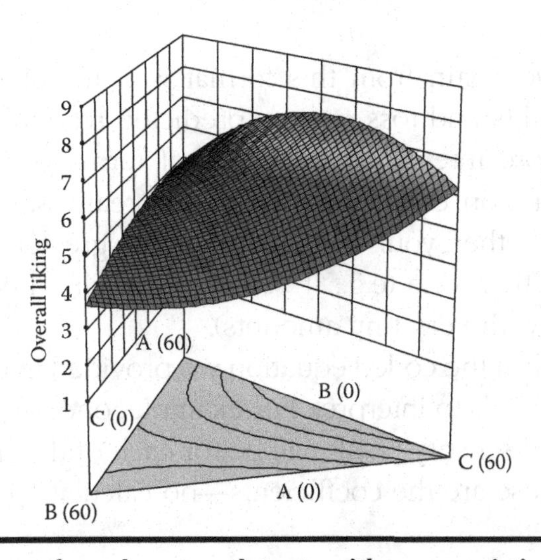

Figure 2.6 Response surface shows peak taste with a synergistic blend of two beers.

negative coefficient (−4.65) on model term BC. You can see this downturn in the response surface along the BC edge. It is less of a deviation from linear blending than is observed for the AC binary blend (BC < AC).

That leaves one coefficient to be interpreted—that of term AB. It turns out that the p-value for the statistical test on this coefficient (2.01) exceeds 0.1, that is, there is more than a ten percent risk that it could truly be zero. (In contrast, the coefficients for terms AC and BC were both significant at the 0.01 level). This time around, we did not bother to exclude the insignificant term (AB) from the model. Removing it would make little difference in the response surface—just a straight edge between the wheat beer (A) and lager (B), rather than a slightly upwards curve. We will revisit the issue of model reduction later. As the number of components increase and modeling gets more complex, it will become cumbersome to retain insignificant terms.

WOULDN'T IT BE EASIER TO MODEL MIXTURES IN ACTUAL UNITS?

Statistical software that can fit Scheffé mixture models may offer these to users in either coded or actual form, or both. In this case, the actual equation for grams of beer is:

$$\begin{aligned}\text{Overall Liking} = {} & 0.082031\ \text{Wheat} + 0.061396\ \text{Lager} + 0.10214\ \text{Black} \\ & + 0.000559187\ \text{Wheat} * \text{Lager} + 0.00204067\ \text{Wheat} * \text{Black} \\ & - 0.00129267\ \text{Lager} * \text{Black}\end{aligned}$$

One advantage you gain from this format is being able to plug in the actual blend weights and toss out the predicted response for overall liking. This gets more intense as the order of terms goes up due to the exponential impact on coefficients—they get really small or very large, depending on whether your actual inputs are greater than one (as in this case) or less than one (e.g., if you were serving beer to ants—they would be happy with very tiny amounts).

Now, look back at the coded equation we provided in the main text and consider how easy it is to interpret. For example, one can see immediately what the predicted sensory result will be for each of the pure components (A, B, and C)—these are the coefficients—no calculating required.

(*Continued*)

So, here's the bottom line: For interpretation purposes, always use the coded equation as your predictive model.

PS: In case you were wondering (?), neither the coded nor the actual equation features a coefficient for the block effect. These models are intended for predicting how an elite beer drinker will react to these three types and their blends. True, some of these individuals will feel compelled to be snobby and look down on all beers, but this cannot be anticipated by the formulator, nor controlled once a product goes up for sale. Thus, the block effect provides no value for predicting future behavior—only to explain what happened during the experiment.

As discussed in Chapter 1, statisticians like to keep models as parsimonious as possible, so let's see if we could get by with only the linear model. Figure 2.7 provides an enlightening view of the BC edge (two-component) after the least-squares fit without the non-linear blending terms (AB, AC, BC).

Notice that all 6 of the actual results (4 of which are at the same level of liking, as noted by the point on the graph) fall below the predicted value

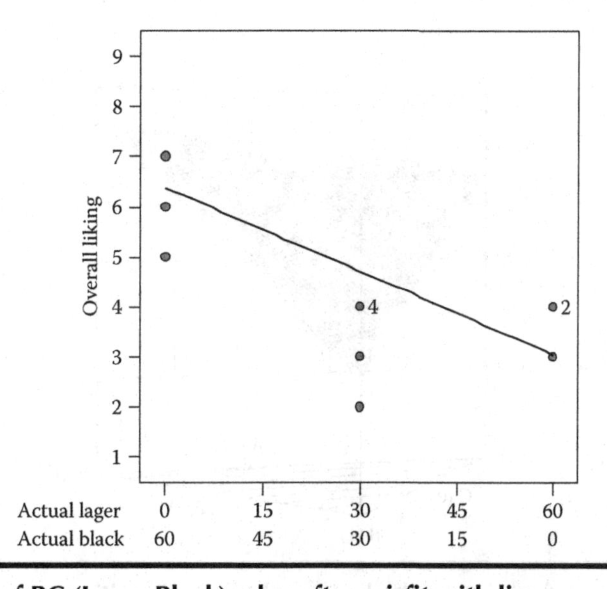

Figure 2.7 View of BC (Lager-Black) edge after misfit with linear model.

from an over-simplified linear blending model. The cumulative impact of these deviations far exceeds the value expected based on the pure error pooled from the four replicate blends tasted by each tester. Thus this linear model exhibits a significant ($p < 0.1$) lack of fit. Clearly the surface needs to dip down at the binary blend of B (lager beer) and C (black), which it does when fit with the quadratic model (look back at Figure 2.5). Not surprisingly, this second-order model for nonlinear blending does not exhibit a significant lack of fit ($p > 0.3$), that is, it fits!

Taking Cost into Account

Given the expense per 12-ounce serving of each of the three beers as detailed at the outset of this case, it was a simple matter mathematically to compute the blended costs shown in the last column of Table 2.1. (The software that we provide for the practice problems will calculate this for you: Simply enter the cost equation.) If one wanted to reduce expense, mixing in a cheap amber-lager like Budweiser (component B) would help as you can see in the response surface on cost in Figure 2.8.

However, in this case, you get what you pay for—the cheaper the blend, the less likable it becomes. You can see this in the side-by-side comparison of contour plots for overall liking versus cost in Figure 2.9a and b. To make it even more obvious, we flagged the optimums of maximum liking versus minimum cost.

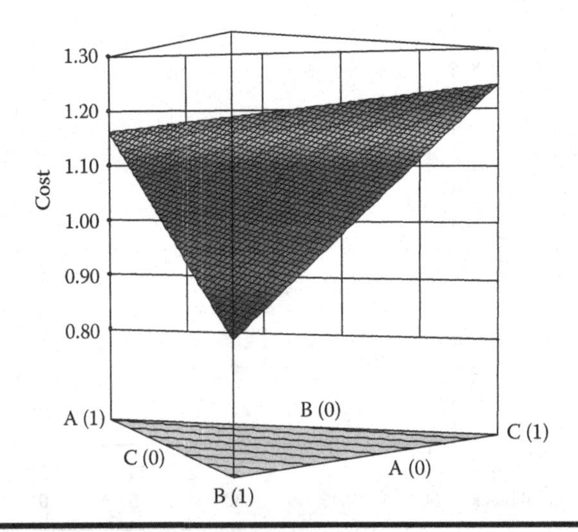

Figure 2.8 Cost as a function of the beer cocktail composition.

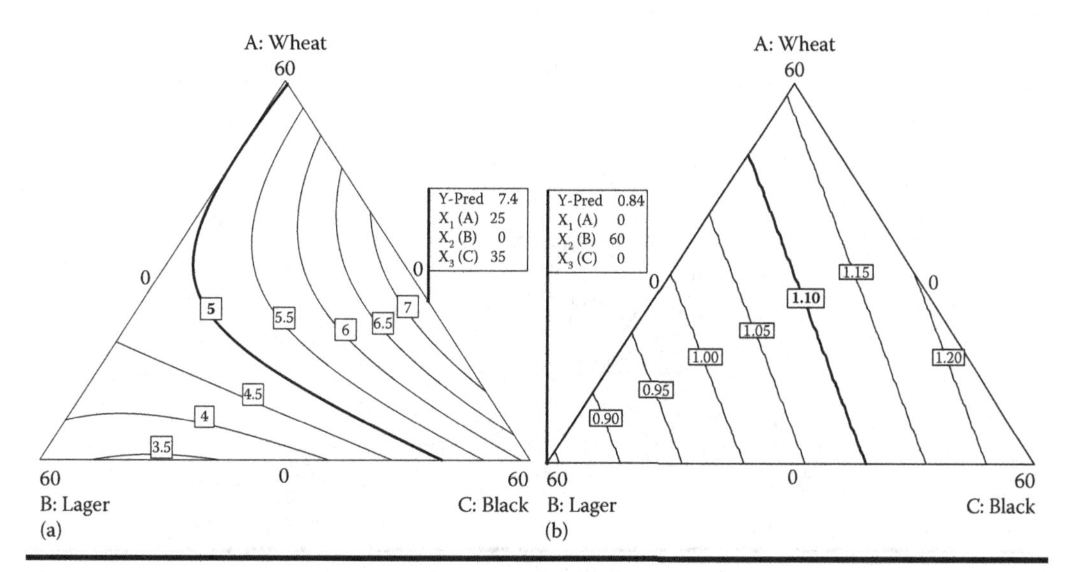

Figure 2.9 (a) Contour plots for overall liking and (b) cost of blended beers.

SEARCHING OUT THE OPTIMUM NUMERICALLY

Figuring out which blend of beer is cheapest is a "no-brainer"—even a six-pack drinker could figure out it's the Budweiser American lager, pure and simple. That's because the cost is a simple linear function. Also, it's deterministic, that is, not derived empirically. However, it would take some calculus to find the optimum of the second-order model for overall liking—perhaps a bit beyond the average beer drinker. As we will discuss later, things get a lot more complicated when searching out the most desirable blends on the basis of multiple responses. It turns out that for computational purposes a hill-climbing algorithm generally works best for this purpose. That's how we came to the flagged optimum of 25 grams of Blue Moon mixed with 35 grams of Black Lager for the (theoretically) most likable Black & Blue Moon cocktail. As a practical matter, Mark blends these two premium beers half and half (no cheap lagers allowed!).

Now suppose as a beer-tender you'd be satisfied to serve a cocktail with an overall liking at or above 5, midway on the hedonic scale of 1 ☹ to 9 ☺. We highlighted this contour in Figure 2.9a. Furthermore, assume that you need to hold the cost to $1.10 per 12-ounce serving—the highlighted contour on Figure 2.9b. Years ago, before the onslaught of presentation programs like Microsoft PowerPoint, statisticians would transfer contour graphs to individual transparencies, shade out the undesirable regions and overlay

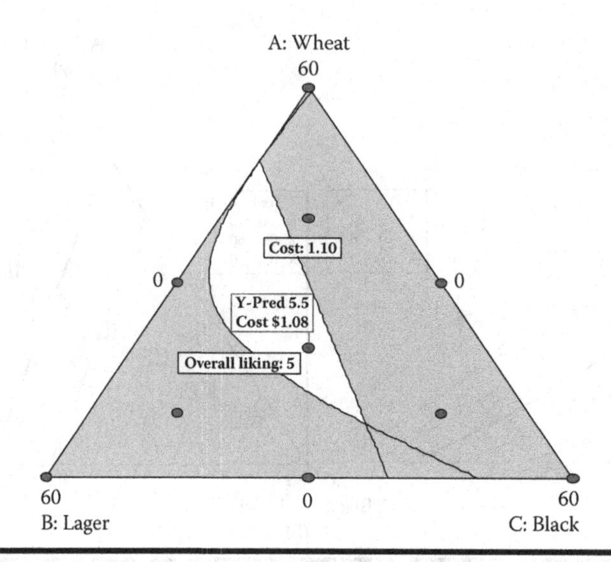

Figure 2.10 Contour plots for overall liking and cost overlaid.

all the transparencies (also known as "view foils") on an overhead projector. Ideally, that left a window of opportunity, or "sweet spot," like that shown in Figure 2.10—produced directly by modern computer software.

The flag marks the centroid blend with equal measures of all three beers, which falls inside the window of overall likability at 5 or higher and cost less than $1.10 per serving. Mark cannot build up much enthusiasm for this—too much work and not first-class. He is an elitist when it comes to the finer things in life, such as a cold beer on a hot evening sitting on the back porch after a long, hard day at the office. Thus, the binary blend of the "Black and Blue Moon" cocktail gets his nod.

Do Not Put a Square Peg into a Triangular Hole

We hope that by now you see the merit of mapping mixtures to the triangular space unless the experiment involves only two components, in which case a simple number line suffices, as illustrated in Chapter 1. Although you are convinced, it may be hard to convert your fellow formulators who remain "square" by sticking to the factorial space used by process developers. Here's a postscript to the beer cocktail case that may help you turn the tide to the triangle.

At this stage, the "Black and Blue Moon" cocktail has achieved a foothold in the drink recipes for those who enjoy a relaxing beverage now and then. However, some dispute remains on the precise proportions of the Black Lager versus the Blue wheat ale. An experimenter who has mastered

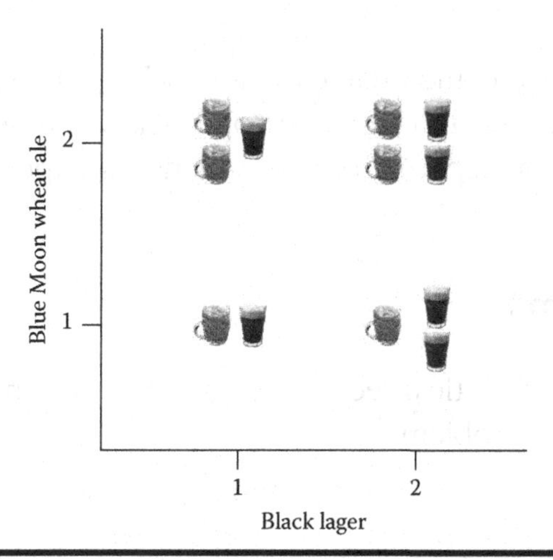

Figure 2.11 Factorial design on formulation of Black and Blue Moon beer cocktail.

two-level factorials, but remains ignorant of mixture design, creates the experiment laid out in Figure 2.11.

If this graph makes you thirsty, then take a brief break from your reading to pour a little something. OK, now concentrate! What's wrong with this picture given that taste is a function of proportions and not amounts of a beverage? The problem occurs along the diagonal running from lower left to upper right in this square experimental region. One of each of the beers versus two of each makes no difference once blended—a sip tastes the same. All that's happening in this direction is a scale-up of the recipe. That would be a waste of good beer!

WHEN YOU KNOW THAT THE BEER DRINKING HAS GOTTEN OUT OF HAND

Mark and Pat are big fans of baseball and their hometown club—the Minnesota Twins. Mark shares season tickets with a relative, but he did not hold seats for a critical year-end game. Thus he was forced to purchase a spot in the outfield bleachers where the really rabid boosters congregate. As he worked his way down the row, Mark observed that the portly fellow in the adjoining seat had lined up 9 beers under foot—one for each inning, thus avoiding the need to fight the lines at the concession stand. To avoid such abuse of alcohol, the policy of the Twins is two beers per purchase,

(*Continued*)

so this guy must have come quite early to stock up so! Such behavior goes beyond the pale of good taste in our opinion. We urge you to be more moderate if you try to replicate our beer-blending experiments.

Practice Problems

To practice using the statistical techniques you learned in Chapter 2, work through the following problems.

Problem 2.1

Normally, we do not recommend the simplex centroid design, especially if done "by the book," that is—without check blends. However, it can be useful for three components that cannot conveniently be broken down too far into fractions. A case in point is the "Teany Beany Experiment" we detailed in *DOE Simplified, 2nd Edition* (Chapter 9). Per a randomized plan, a dozen or so tasters were each required to chew one or more small jelly beans flavored with apple, cinnamon, or lemon. Biting down on all three at once presented a challenge, but everyone got the job done. Table 2.2 shows the experiment design and the taste ratings.

Table 2.2 Teeny-Beany Mixture Design—A Three-Component Simplex Centroid

ID#	*Blend*	*A: Apple (%)*	*B: Cinnamon (%)*	*C: Lemon (%)*	*Taste Rating*
1	Pure	100	0	0	5.1
2	"	"	"	"	5.2
3	Pure	0	100	0	6.5
4	"	"	"	"	7.0
5	Pure	0	0	100	4.0
6	"	"	"	"	4.5
7	Binary	50	50	0	6.9
8	Binary	50	0	50	2.8
9	Binary	0	50	50	3.5
10	Centroid	33.3...	33.3...	33.3...	4.2
11	"	"	"	"	4.3

This table translates the relative proportions to a percent scale. Beware! Software that supports mixture design may require you to make each row add to the specified total, in this case, one hundred. If so, blend number 10 will fail unless you "plug" it by entering 33.34 for one of the three components. (A reminder: Statistical software used to do the computations can be freely accessed via a website developed in support of this book. There you will also find answers posted. See *About the Software* for instructions).

Note from the response column that each of the three pure-component blends was replicated, as well as the overall centroid (the three "beaner"). The results come from the averaging of several tasters, done for simplicity sake. Analyze this data. What hits the sweet spot for this candy-loving crowd?

MORE SOPHISTICATED APPROACHES FOR TASTING TINY JELLY BEANS

To keep things simple, we provide only the average taste rating for this case study on candy. However, a more precise analysis blocked out the rating shifts caused by personal preferences—some scoring high across the board (people with a "sweet tooth") and others low. We found a remarkable consistency—all but one of the tasters liked cinnamon best and lemon least. However, one individual loved lemon far and away. We excluded her from the average. But do not fear, she remains free to fill her candy jar with these yellow beans.

I'm not an outlier; I just haven't found my distribution yet.

—Ronan M. Conroy

Problem 2.2

This is a case where materials can be freely blended, thus the formulators could augment the simplex centroid with check blends. The experimenters measured the effects of three solvents known to dissolve a particular family of complex organic chemicals (Del Vecchio, 1997). They had previously discovered a new compound in this family. It needed to be dissolved for purification purposes, so they needed to find the optimal blend of solvents.

Table 2.3 Design Matrix and Data for Solvent Study

ID#	A *MEK*	B *Toluene*	C *Hexane*	*Blend Type*	*Solubility (g/L)*
1	100	0	0	Pure A	121
2	0	100	0	Pure B	164
3	0	0	100	Pure C	179
4	50	50	0	Binary AB	140
5	0	50	50	Binary BC	180
6	50	0	50	Binary AC	185
7	33.3	33.3	33.4	Centroid	199
8	66.6	16.7	16.7	Check	175
9	16.7	66.6	16.7	Check	186
10	16.7	16.7	66.6	Check	201

Table 2.3 shows the experimental design and results. Remember that the actual run order for experiments like this should always be randomized to counteract any time-related effects due to the aging of material, and so on. Also, we recommend that you always replicate at least four blends to get a measure of pure error. In this case, it would have been helpful to do two each of the pure materials and, also, replicate the centroid.

Notice that for the sake of formulating convenience, the interior points (centroid and check blends) were rounded to the nearest tenth of a percent so that they always added to one hundred. Which chemical, or a blend of two or three, will work best as a solvent? (Hint: Read the Appendix before finalizing your answer.) For extra credit on this problem, determine the relative costs of each chemical (MEK is methyl ethyl ketone) and work the material expense into your choice.

Appendix 2A: The Special Cubic (and Advice on Interpreting Coefficients)

In the Appendix to Chapter 1 we detailed a very sophisticated third-degree mixture model called the "full cubic." As we illustrated, this equation captures wavy response behavior that would be unusual to see in a typical formulation experiment. As a practical matter (remember the principle of parsimony) it often suffices to apply a simpler "special cubic" model, such as this one for three components:

$$\hat{y} = \beta_1 x_1 + \beta_2 x_2 + \beta_3 x_3 + \beta_{12} x_1 x_2 + \beta_{13} x_1 x_3 + \beta_{23} x_2 x_3 + \beta_{123} x_1 x_2 x_3$$

As you may've already surmised, this modeling option is not applicable to two components. For three components, the special form saves three model terms over the full cubic, and it's easier to interpret. Figure 2A.1 provides an example of three components of a solid rocket fuel blended for a peak in the measured response—elasticity (Cornell and Piepel, 2008, p. 55).

Notice the dramatic upward curve along the BC edges—a clear case of synergism between these two materials (oxidizer and fuel). However, a further increase occurs with the addition of component A (binder)—a move from the center of the BC edge to the flagged peak in the interior of the triangular mixture space. This represents a beneficial non-linear blending involving three components.

The fitted special cubic model (coded) that generated this response surface is:

$$\text{Elasticity} = 2351A + 2446B + 2653C - 6AB + 1008AC + 1597BC + 6141ABC$$

The coefficient on ABC seems surprisingly large unless you remember that the components are on a scale of 0 to 1. For example, recall from the last chapter that the maximum deflection from linear blending occurs at the

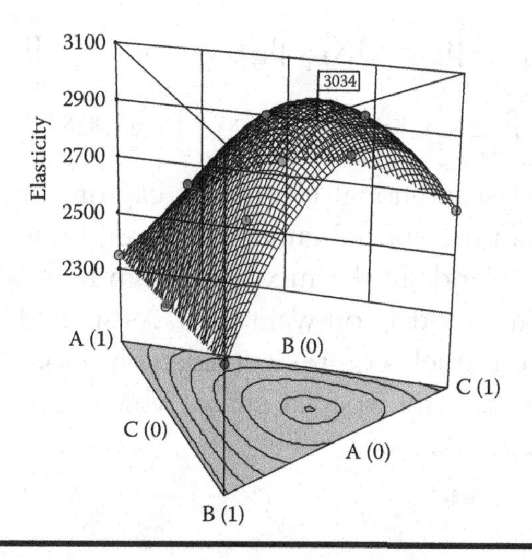

Figure 2A.1 Response surface from special cubic model.

1/2–1/2 ("50/50") point, thus you must multiply the two-component terms, such as BC, by one-fourth (1/4) to assess the synergism (or antagonism). However, for the three-component term, the maximum non-linear effect occurs at the 1/3-1/3-1/3 point (centroid) with a magnitude of 1/27 the coefficient of that term. Thus, in this case the maximum effect for BC of approximately 400 (1/4 of 1597) almost doubles the greatest impact of ABC (1/27 of 6141).

How do higher order terms compare to the linear ones in this case (2351, 2446, and 2653)? Here again, you must be careful not to jump to conclusions without first considering the meaning of linear coefficients in Scheffé polynomial mixture models—the difference is what counts, not the absolute magnitude. The range of linear coefficients is only a bit over 300 (2653 for C minus 2351 for A), so the tilt in the plane of response (upwards to component C) is actually exceeded by the synergism of B and C.

Have you had enough of trying to interpret coefficients in these higher order mixture models? We hope so because it's not worth belaboring—simply look at the response surfaces to get a feel for things. Then with the aid of computer tools, use the model to numerically pinpoint the most desirable recipe for your formulation.

A FURTHER WRINKLE: THE SPECIAL QUARTIC

Another variation on mixture models is the "special quartic"—shown here for three components (Smith, 2005):

$$\hat{y} = \beta_1 x_1 + \beta_2 x_2 + \beta_3 x_3 + \beta_{12} x_1 x_2 + \beta_{13} x_1 x_3 + \beta_{23} x_2 x_3$$
$$+ \beta_{1123} x_1^2 x_2 x_3 + \beta_{1223} x_1 x_2^2 x_3 + \beta_{1233} x_1 x_2 x_3^2$$

This model provides additional terms that capture more complex non-linear blending than the special cubic. However, be forewarned that the number of unique blends in the mixture design must always exceed the count of terms in the model you want to fit. As spelled out in Table 2A.1, the special quartic model requires considerably more blends at four or more components than the special cubic, which may make the experiment unaffordable.

(Continued)

Table 2A.1 Number of Terms in Full versus Special Scheffé Polynomials

Components (q)	*Linear*	*Quadratic*	*Special Cubic*	*Full Cubic*	*Special Quartic*	*Full Quartic*
2	2	3	NA	4	NA	5
3	3	6	7	10	9	15
4	4	10	14	20	22	35
5	5	15	25	35	45	70

Chapter 3

Simplex Lattice Designs to Any Degree You Like

> My beloved is like a gazelle or a young stag. Behold, he is standing behind our wall!
> He is looking through the windows. He is peering through the lattice.
>
> **—Song of Solomon 2:9**

In this chapter, we will detail the design of lattices within a simplex geometry. The procedure will work for any number of components, but we will illustrate simplex lattice design only up to four ingredients, which form a three-dimensional tetrahedron. That will provide enough of a challenge for now.

Working with Four Components in Tetrahedral Space

To get a feel for a tetrahedron, make one out of a piece of paper (preferably ruled for graphs) per the following procedure (Box and Draper, 2007, p. 512). Start by drawing an equilateral triangle as best as you can (perfectionists see the sidebar). Then using the midpoints of the edges as vertices, inscribe a second equilateral triangle as pictured in Figure 3.1a.

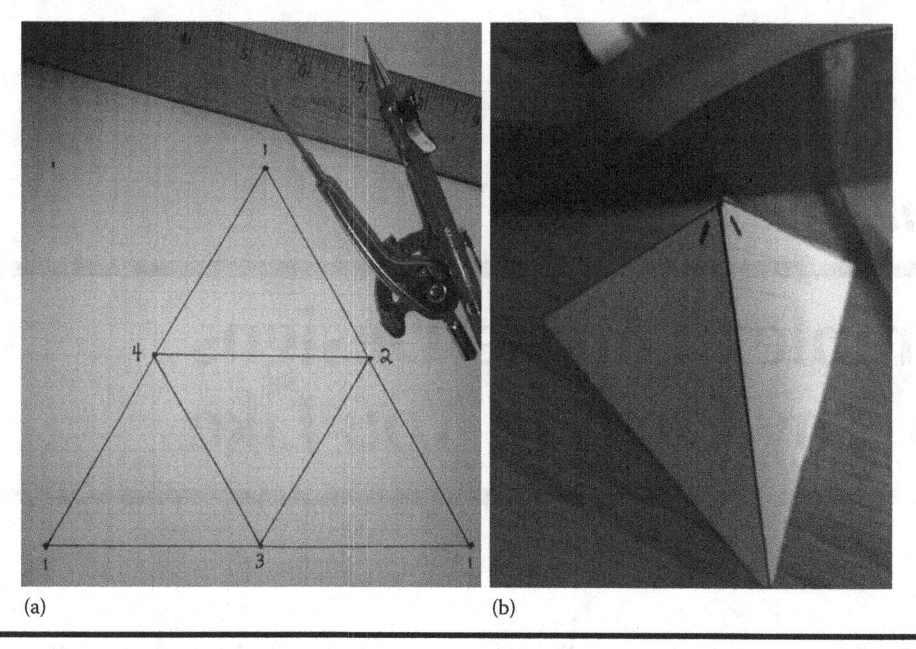

(a) (b)

Figure 3.1 (a) Laying out the fold lines for building a tetrahedron and (b) the resulting 3D paper model.

Label the vertices of the larger triangle "1"—these represent the first ingredient. Identify the corners of the smaller triangle as 2, 3, and 4 to pinpoint three more components (thus allowing for four, total). Now cut out the large triangle and fold the 1's along lines 4–2, 2–3, and 3–4 to a point, as shown in Figure 3.1b. There—you've made a tetrahedron! Keep this handy to help you visualize our illustrations of four-component mixture design.

HOW TO DRAW AN EQUILATERAL TRIANGLE

It's easy to draw a triangle, but surprisingly hard to make it equilateral, that is, with sides of precisely equal length. The trick is to start with the proper tool—a compass—and knowing how to use it for drawing arcs and setting fixed distances. We could detail this here, but it will make a lot more sense for you to see it demonstrated by The Math Open Reference Project at the internet web page for "Constructing an Equilateral Triangle": www.mathopenref.com/constequilateral.html.

Building a Simplex Lattice Design

Henry Scheffé's pioneering publication "Experiments with Mixtures" (noted in a Chapter 1 sidebar) introduced simplex lattice design to industrial statisticians and their client formulators. The name of this mixture design provides two big clues about its construction:

- A *simplex* is a geometric figure having a number of vertexes or corners equal to one more than the number of dimensions of the variable space for n dimensions. For example, when n equals two, the simplex has three corners—an equilateral triangle.
- A *lattice* is an arrangement in space of isolated points in a regular pattern, such as atoms in a crystalline solid.

The simplex lattice design comprises m+1 equally spaced values from 0 to 1, thus

$$x_i = 0, 1/m, 2/m, \ldots, 1$$

where x is the individual (i) component in coded levels and m represents the degree of the polynomial that the formulator feels will be needed to fit the experimental response. The number of blends in a simplex lattice design depends on both the number of components (q) and the degree of the polynomial. The simplest (overly so!) simplex lattice is this one designed for two components at a degree of one (m = 1):

$$x_i = 0, 1/m, \cancel{2/m, \ldots}$$

$$x_i = 0, 1/1$$

$$x_i = 0, 1$$

There are only two points, 0 and 1! Going beyond 1 is not allowed, so the design must stop there. It's designated as "(2, 1)" based on the number of components and degree; respectively. We do not recommend this (2, 1) simplex lattice as-is, there being too few points for any appreciable power. However, it serves as a launching pad to designs for three components that are not that bad – for example, the two pictured in Figure 3.2a and b for second and third-degree modeling.

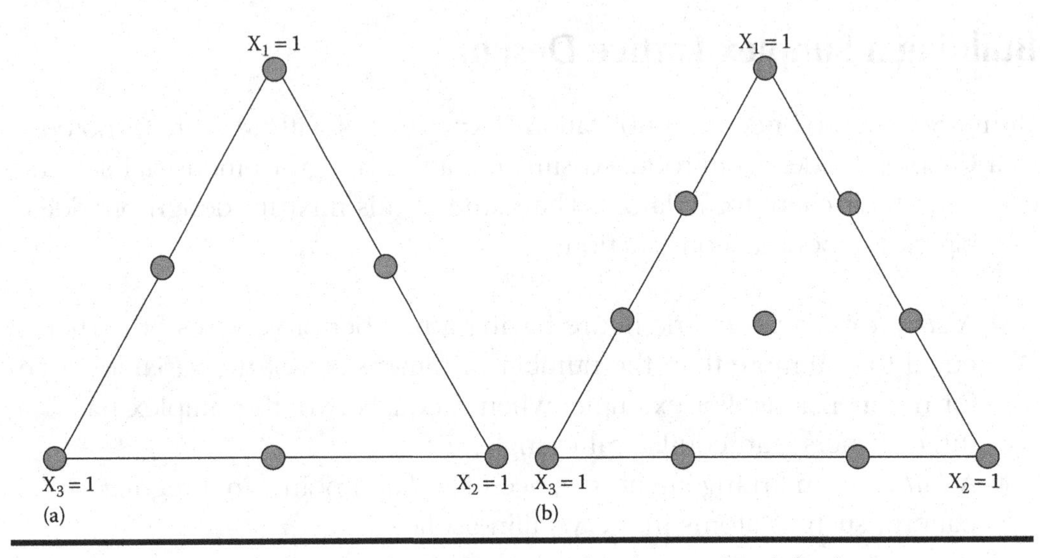

Figure 3.2 Three-component simplex lattices of 2nd (a) and 3rd (b) degree.

If you are a formulator, it will seem odd to experiment with several components but never a complete blend, which is exactly what happens with the (3, 2) design depicted in Figure 3.2a—its interior remains devoid of points. However, keep in mind that you need not adhere strictly to this template—strongly consider adding the centroid and, if not impractical, additional check blends inside the simplex formulation space. This will be demonstrated by example later in this chapter.

Two more simplex lattice designs are shown in Figure 3.3a and b. Notice by their geometry—tetrahedral—that these encompass four components.

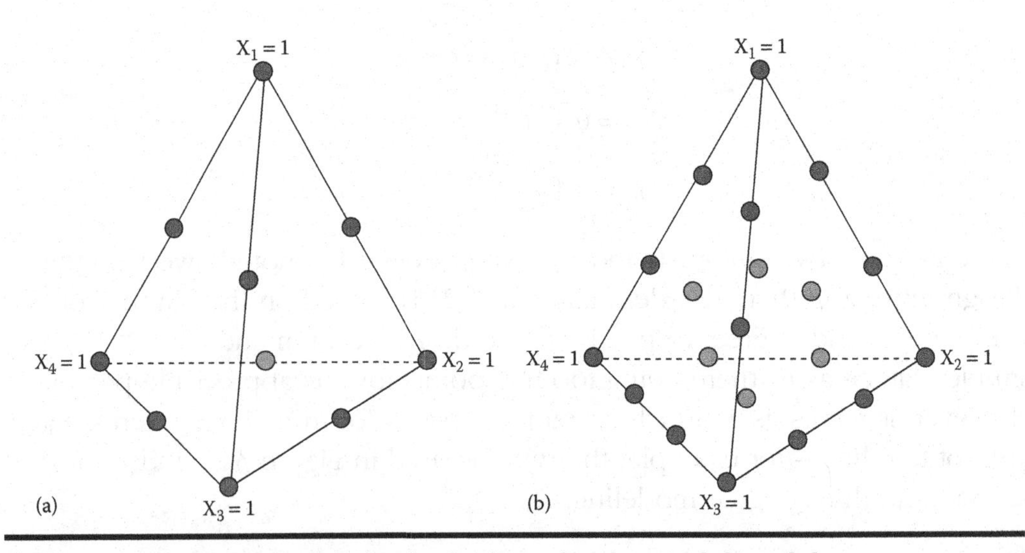

Figure 3.3 Four-component simplex lattices of 2nd (a) and (b) 3rd degree.

Also, evident at-a-glance is the increase in the degree of the lattice from left to right of two to three; respectively.

An easy way to infer the degree is by the number of design points along the edges; when broken in half, the degree is two—whereas a fragmentation by thirds indicates a third-degree lattice.

AN EXCITING (!) FORMULA FOR THE NUMBER OF DESIGN POINTS IN A SIMPLEX LATTICE

The number of design points N in a simplex lattice depends only on the number of components q and the degree of polynomial m. The equation is:

$$N = (q + m - 1)!\ /(m!\ (q - 1)!)$$

You may recall from math and/or stats class that the exclamation marks denote a factorial notation. This will come back to you more quickly by following this example calculation on the number of design points for a four-component simplex lattice designed to the third degree—a (q, m) of (4,3), which computes as:

$$= (4+3-1)!/(3!(4-1)!) = (6!)/(3!3!) = (6\times5\times4\times\cancel{3\times2\times1})/[(3\times2\times1)(\cancel{3\times2\times1})]$$

Notice that the multiplication of 3 × 2 × 1 appears both in the numerator (top) and denominator (bottom) of this equation. Thus, these numbers cancel, and the equation simplifies to:

$$N = (6\times5\times4)/(3\times2\times1) = (\cancel{6}\times5\times4)/\cancel{6} = 5\times4 = 20 \text{ points}$$

Check this calculation by counting the number of points in Figure 3.3b, comprised of 4 pure blends at the four vertices, 12 two-part blends along the six edges and 4 three-part blends at the centers of the four faces. That comes to 20. Yes?

PS: For purposes of building simplex lattice designs, the special model forms are considered to be full degree. So, for example, the degree (m) for the special cubic is 3—the same as it would be for the full cubic Scheffé polynomial.

Augmented Simplex Lattice: When in Doubt, Build Them Stout

As we've shown, the simplex lattice design may not provide a complete mixture—in fact, this glaring deficiency occurs whenever the number of components exceeds the degree ($q > m$). Thus the first step for a practical augmentation will be to add the overall centroid—a complete blend with equal amounts ($1/q$) of each component. However, there's a less obvious drawback to running the raw simplex lattice as spelled out by textbook—they often use up all the degrees of freedom for model-fitting, with none left over to provide check blends for lack of fit test. For example, notice how in Table 3.1 the number of points in the second-degree ($m = 2$) simplex-lattice designs equals the number of terms in the quadratic mixture model (second-order Scheffé polynomial). These are said to be "saturated" designs. If the formulator can be certain of the degree, a saturated experiment would be most efficient. But that level of knowledge usually comes after all the research, not before. Otherwise, why bother experimenting?

In case you may be tempted to fall back on the less sophisticated simplex centroid, which does guarantee inclusion of a complete blend (by definition), notice how the number of points increases exponentially—going far beyond the minimum required for the quadratic mixture model. Thus, we advise the simplex centroid only for exceptional cases for three components only, such as the taste-testing experiment presented in Problem 2.1.

Instead, we recommend an augmented simplex lattice (ASL) that incorporates these added points:

- 1 centroid (if needed)
- q axial check blends
- At least 3 replicates

Table 3.1 Number of Points in Textbook (Raw) Simplex Designs versus What's Required by Model

Components (q)	*Simplex Centroid*	*Simplex Lattice (m = 2)*	*Quadratic Model Terms*
3	7	6	6
4	15	10	10
5	31	15	15
6	63	21	21

We detailed axial check blends via the beer blending case study presented in Chapter 2, so you're aware of how these points fill the gaps between the centroid and each of the q vertices. This design also featured replicates and how they provide measures of pure error, which, in conjunction with check blends, facilitate testing for lack of fit. Now we are establishing this augmentation as the standard procedure for simplex designs—centroid or lattice.

HOW MANY POINTS TO REPLICATE AND WHICH ONES

We're trying to be sensible in advising how many points need to be replicated. Having none provides zero degrees of freedom (df) for estimating pure error, so adding one replicate is infinitely better from a mathematical perspective (one divided by zero). However, when computing the lack-of-fit test, a very large ratio of variance from the benchmark pure error would be required. For example, going back to the gold-copper mixture experiment in Chapter 1, here are the critical values of this ratio at a p-value of 0.05 for df's of 1, 2, 3, 4, and 5, respectively: 199.5, 19.0, 9.5, 6.9, and 5.8.* Notice how quickly this value falls but then it starts leveling off after 3 df. By 5 df the critical F becomes stable. Thus, we recommend that at least five points be replicated in any mixture design.

Now which points will be best to replicate? We suggest picking points that exhibit the highest leverage. You may be familiar with this statistical term: It's simply a measure of potential influence for any given point. Consider again, for example, the gold-copper blending experiment. If the goldsmith only checked the melt point on one sample each of pure gold and pure copper, these two results would be entirely influential on a linear model, that is, both design points exhibit a leverage of one. However, replicating these points—increasing the experimental runs from two to four—reduces the leverage to 0.5 for each.

Computing leverage becomes more complicated for higher-order models,** but let me give you some feel for how it works—Here are the measures for the gold–copper experiment without any replicates: 0.89 at the vertices (pure metal), 0.49 at the centroid ("50-50" gold-copper) and 0.37 for the axial check blends (1/4th-3/4th and 3/4th-1/4th). The three highest leverage points are the two vertices and the centroid, so these were the ones chosen for replication.

(Continued)

*(From 5% table provided in Appendix 1–3 of DOE Simplified with 2 df in the numerator of the F-ratio—starting from the top of the second column and looking down from 1 df in the denominator—the first row.)

**(For more details on leverage, see the sidebar titled "Assessing the Potential Influence of Input Data via Statistical Leverage" provided in Chapter 1 of *RSM Simplified*.)

Second-order designs augmented per the guidelines we've provided are suitable for producing a "simplex response-surface" (Smith, p. 49). This leads to an important insight: Mixture design for an optimal formulation is a close cousin to response surface methods (RSM) for process optimization.

Using Augmented Simplex Lattice Mixture Design to Optimize an Extra-Virgin Olive Oil

Olive oil, an essential commodity of the Mediterranean region and the main ingredient of their world-renowned diet (see sidebar), must meet stringent European regulatory guidelines to achieve the coveted status of "extra virgin." Oils made from single cultivars (a particular cultivated variety of the olive tree) will at times fall into the lower "virgin" category due to seasonal variation. Then it becomes advantageous to blend in one or more superior oils based on a mixture design for optimal formulation. For example, a team of formulators (Vojnovic et al., 1995) experimented on four Croatian olive oils—Buza, Bianchera (pronounced "be an kay ra"), Leccino (pronounced "la chee no") and Karbonaca—to achieve an overall sensory rating of at least 6.5 on a 9-point hedonic scale, thus easily exceeding the cut off for "extra virgin" (5.5 considered to be "virgin"). The ratings were done by ten assessors trained on fundamental tastes (sweet, salt, sour, and bitter) and defects of virgin olive oils, such as rancidity (Figure 3.4).

We've adapted a bit of the original study, simplified it, and made it more educational while capturing the essence of how these formulators made use of mixture design and what they discovered.

Thus the assessors can discern very tiny differences in sensory attributes that may depend on the subtle nonlinear blending of two or more oils. Therefore, the four-component ($q = 4$) simplex lattice is set up to the third degree ($m = 3$). That produces 20 unique blends as shown in Figure 3.3b.

Figure 3.4 Olive oil. (Royalty-free Internet post 670878 by Stockxpert.)

To augment this lattice, the formulators add 4 axial check blends to the overall centroid. They then specify that the four vertices (chosen for their high leverage) and centroid be replicated (for added pure error measure) at random intervals. (Always randomize!) Assume that the formulators use a 1-liter blender to mix the oils—30 blends in total after the augmentation. This ASL design and the results for overall sensory ratings are shown in Table 3.2. (Note that, for the

Table 3.2 ASL Design for Blending Four Olive Oils and Their Sensory Results

#	*Point Type*	*A: Buza*	*B: Bianchera*	*C: Leccino*	*D: Karbonaca*	*Sensory Rating*
1	Vertex	1	0	0	0	6.98
2	"	1	0	0	0	6.84
3	Vertex	0	1	0	0	6.49
4	"	0	1	0	0	6.45
5	Vertex	0	0	1	0	7.25
6	"	0	0	1	0	7.30
7	Vertex	0	0	0	1	5.88
8	"	0	0	0	1	5.95
9	Third edge	0.667	0.333	0	0	7.38
10	Third edge	0.333	0.667	0	0	7.12

(*Continued*)

Table 3.2 (*Continued*) ASL Design for Blending Four Olive Oils and Their Sensory Results

#	*Point Type*	*A: Buza*	*B: Bianchera*	*C: Leccino*	*D: Karbonaca*	*Sensory Rating*
11	Third edge	0.667	0	0.333	0	6.87
12	Third edge	0	0.667	0.333	0	6.84
13	Third edge	0.333	0	0.667	0	6.95
14	Third edge	0	0.333	0.667	0	7.17
15	Third edge	0.667	0	0	0.333	7.36
16	Third edge	0	0.667	0	0.333	7.14
17	Third edge	0	0	0.667	0.333	7.50
18	Third edge	0.333	0	0	0.667	7.16
19	Third edge	0	0.333	0	0.667	6.95
20	Third edge	0	0	0.333	0.667	7.00
21	Triple blend	0.333	0.333	0.333	0	7.56
22	Triple blend	0.333	0.333	0	0.333	7.53
23	Triple blend	0.333	0	0.333	0.333	7.29
24	Triple blend	0	0.333	0.333	0.333	7.28
25	Axial CB	0.625	0.125	0.125	0.125	7.41
26	Axial CB	0.125	0.625	0.125	0.125	7.37
27	Axial CB	0.125	0.125	0.625	0.125	7.50
28	Axial CB	0.125	0.125	0.125	0.625	7.19
29	Centroid	0.25	0.25	0.25	0.25	7.58
30	"	0.25	0.25	0.25	0.25	7.55

sake of simplicity, the one-third and two-thirds levels are rounded to 0.333 and 0.667, respectively—thus adding to the total of 1.)

The statistical analysis of this data is detailed via Problem 3.2. The chosen model is a reduced special cubic:

$$\text{Sensory Rating} = 6.91\ A + 6.47\ B + 7.29\ C + 5.93\ D + 2.51\ AB - 0.91\ AC$$
$$+ 3.70\ AD + 0.54\ BC + 3.75\ BD + 2.78\ CD + 11.65\ ABC$$

The presence of the ABC nonlinear blending term supports the choice of a third-degree lattice design. The other three special-cubic terms (ABD, ACD, and BCD) were insignificant ($p > 0.1$) so we chose to remove them from the final model. Rather than laboriously dissecting the model by its remaining terms, let's focus on the response surface graphics: The pictures will tell the story.

Unfortunately, now that we've gone to the third dimension the imaging gets trickier—for example, only three out of the four components can be depicted on a contour plot. This complication provides the perfect opportunity to present the "trace" plot—a way to view the relative effects of any number of components. A trace plot for the olive-oil mixture experiment is shown in Figure 3.5.

The traces are drawn from the overall centroid—all components at equal volume within the 1-liter vessel. This is called the "reference blend." Each component alone is then mathematically varied while holding all others in constant proportion. This reveals, for example, that the predicted sensory evaluation falls off dramatically as the Karbonaca oil (D) is increased relative to the three alternatives.

To give you a better feel of how the trace plot is produced, let's consider the simpler case of only three components. Figure 3.6 shows the paths of the three traces.

The trace for x_1 starts at the overall centroid where it amounts to one-third of the three-component blend. The other two components are

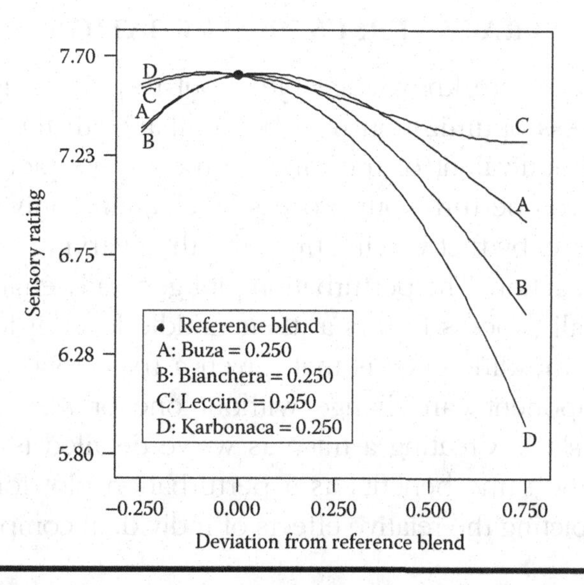

Figure 3.5 Trace plot for olive-oil mixture experiment.

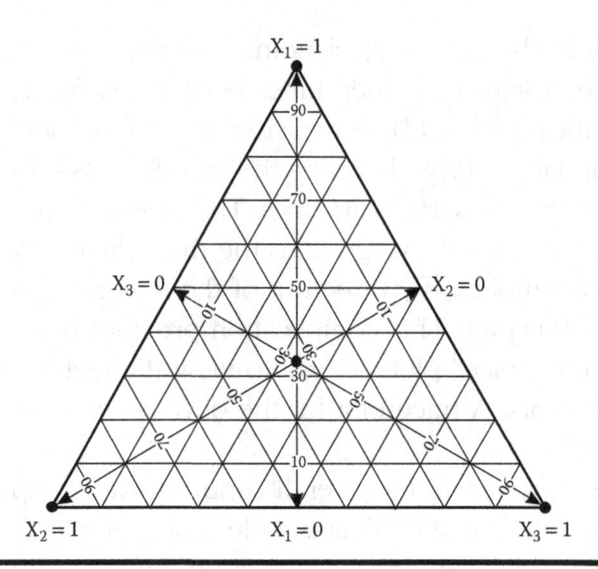

Figure 3.6 Traces for three components only.

also at one-third. Thus their ratio is one-to-one. Tracing x_1 from the centroid down to the base of the ternary diagram reduces the amount of this individual component to zero. At this point the amounts of the other two components become one half each—thus their ratio remains one-to-one. In fact, if you pick any point along any of the three traces, the other two components remain at constant proportions! Try working this out for yourself—it will be good practice for reading off coordinates on the ternary diagram.

TRACE VERSUS PERTURBATION

Those of you who are knowledgeable about response surface methods (RSM) for process optimization may be familiar with the "perturbation" plot—the RSM equivalent of the mixture trace. This plot predicts what will happen if you perturb your process by changing only one factor at a time, for example, by first varying time and then varying the temperature of a chemical reaction. The perturbation plot generally emanates from the center point—all process factors at their middle level. It looks the same as a trace, but the difference is that, given a fixed total of the amount, no single component can change without one or more of the others taking up the slack. Creating a trace as we've detailed is a workaround that provides the same benefits as a perturbation plot for RSM, that is, graphically depicting the relative effects of individual components as they

(Continued)

become more, or less, concentrated in your formulation. However, keep in mind the one-dimensional nature of the trace, which cannot substitute for contour plots or 3D views of the surface as a function of any two components. Only then will you see an accurate picture of non-linear blending effects.

PS Warning: The trace (and perturbation) plot can change dramatically with a change in the reference point. Suggestion: Once you settle on the maximum draw of the trace (or perturbation) plot from that location, it provides perspective on how robust the solution may be for undesirable changes caused by variations in the inputs.

Now that we've been provided with clues on the non-linear blending behavior of the four olive oils, it seems sensible to study the response surfaces of the three good components "sliced" at varying levels of the inferior fourth component. For example, Figure 3.7a and b show the sensory results at the overall centroid (all components, including D, at 0.25 concentration) versus no Karbonaca oil (D = 0). If anything, it's the Bianchera (B) oil that creates the greatest effect on taste—very noticeably on these slices with D held fixed at two specific levels (0 and 0.25).

These response surface graphs are very illuminating! It appears as if the complete four-part blend at the centroid, shown on the left (Figure 3.7a), will be most robust to variations in olive oil concentrations and deliver a

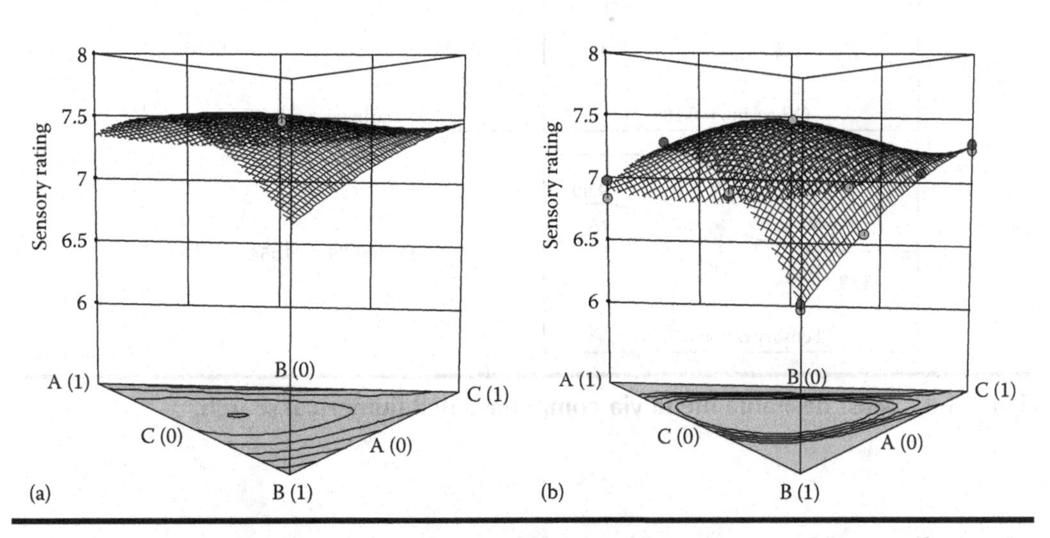

Figure 3.7 (a, b) Sensory results at the overall centroid versus no Karbonaca oil (D = 0).

superior sensory rating for the most part. A more comprehensive computer-aided search of the entire tetrahedral formulation space produced the blend detailed below and depicted in Figure 3.8:

1. 0.333 Buza
2. 0.299 Bianchera
3. 0.189 Leccino
4. 0.179 Karbonaca

This is predicted to produce a sensory rating of 7.63—higher than any of the actual test results. However, any such prediction must be subject to verification.

The upward ramp for the sensory response shows how a rating below 6.5 will be completely undesirable, whereas a rating of 9 represents the peak of desirability (prima!). The predicted optimum falls 0.454 up the scale from zero to one on desirability. This may be the best the blenders can do

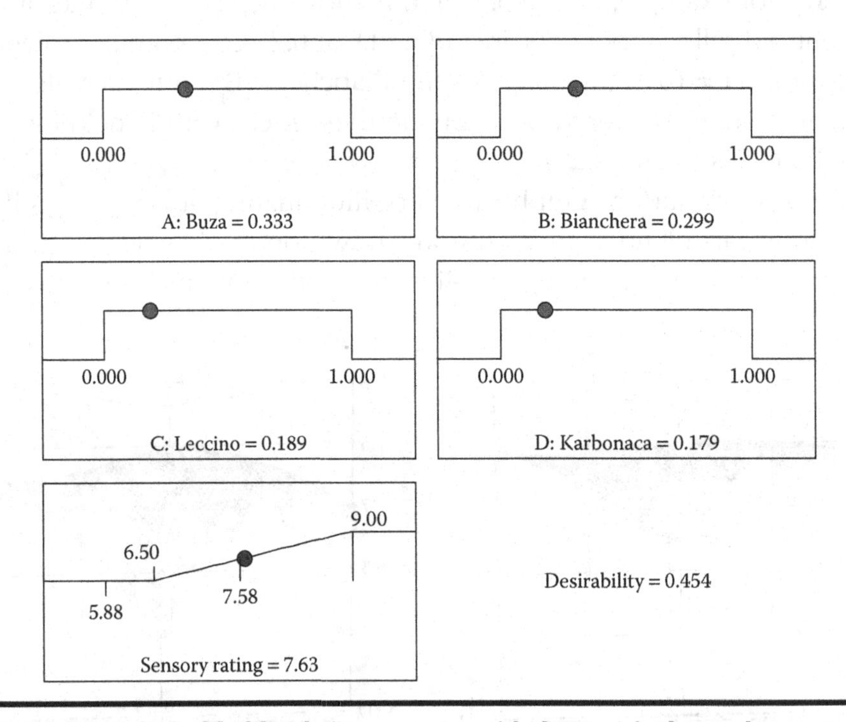

Figure 3.8 Most desirable blend via computer-aided numerical search.

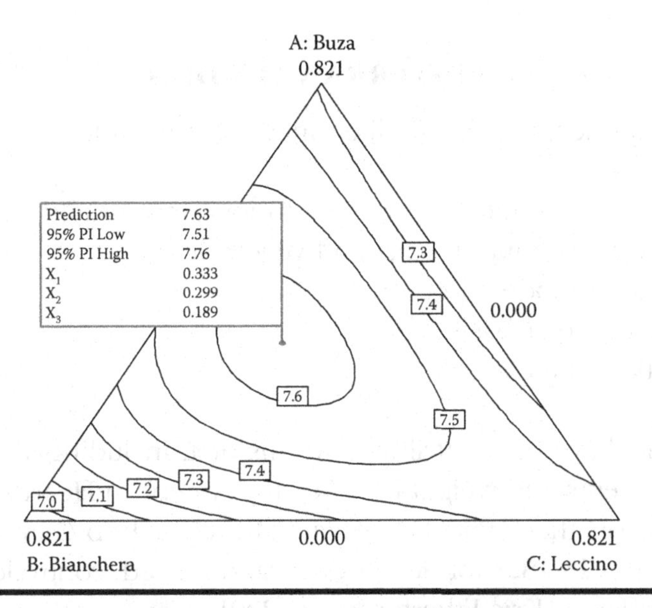

Figure 3.9 Most desirable blend flagged on contour plot (D sliced at 0.179).

with these particular varietals. A sensory outcome of 9 on the hedonic scale remains in the province of the gods who enjoy only the best of the best. Figure 3.9 flags the attainable optimal blend of these four olive oils. It displays the prediction interval between 7.51 and 7.76, based on 95% confidence. Ideally, the verification blend will be rated by the sensory panel within this range. A result outside of the prediction interval would cast doubt on the validity of the model.

Remember that these plots are derived from the Scheffé-polynomial predictive-model fitted to the actual experimental data and validated statistically. However, only by producing an independent confirmatory blend and subjecting it to sensory evaluation will this be verified for all practical purposes.

The optimum sensory rating comes from a blend that goes light on the Karbonaca oil. But what if this inferior oil can be bought very cheaply? Then a blend of it being maximized could be made at a sensory rating just good enough to pass at the extra-virgin level. This might be worth a try!

MEDITERRANEAN DIET

The key components of the Mediterranean diet include:

- Olive oil as an important source of monounsaturated fat
- A generous amount of fruits and vegetables
- Red wine in moderation
- Fish on a regular basis
- Very little red meat

Many benefits have been attributed to this diet, including a reduced rate of coronary events and weight loss. See the American Heart Association's internet post on the "Lyon Diet Heart Study" at http://circ.ahajournals.org/content/103/13/1823 for details on a randomized, controlled trial with free-living subjects (Kris-Etherton et al., 2001).

> Two themes characterize people who have lost a significant amount of weight and kept it off long-term: (1) they don't eat as much as in the past, and (2) they exercise more. Look for these clues when you search for effective weight-loss programs.
>
> **—Dr. Steve Parker, author of *The Advanced Mediterranean Diet* (Vanguard Press, 2008)**

By employing mixture design, these formulators learned how to blend various olive oils to achieve similar qualities. This allowed them to work around limited supplies of specific varieties and adapt their blends to prevailing preferences in the market.

Practice Problems

To practice using the statistical techniques you learned in Chapter 3, work through the following problems.

Problem 3.1

Experience how easy it will be to design a mixture experiment and analyze the results by using a computer tool specialized for this purpose. It can be freely accessed via the web site developed in support of this book: See *About the Software* for the path. When you arrive at the internet page, follow the link to the accompanying tutorials. Then download and print the *Mixture Design Tutorial* (*Part 1—The Basics*). It details a case study on a detergent product that introduces some very practical aspects on how to experiment on only a portion of an entire formulation. If you come across a few new concepts, follow the advice in the *Introduction* of the book, just keep moving, and worry later about the explanations—these will be forthcoming. The primary purpose of this exercise is to get a feel for how dedicated DOE software can facilitate your design and analysis of mixture experiments.

Problem 3.2

Via the web site developed in support of this book, go to the answer posted for this problem—a follow up on the olive oil mixture experiment. It details all the important statistical models and validation of the chosen one by analysis of variance (ANOVA) and residual diagnostics. Your assignment will be to reproduce these results using the software made available to you for this purpose.

Chapter 4

Mixture Constraints That Keep Recipes Reasonable

> The more constraints one imposes, the more one frees one's self. And the arbitrariness of the constraint serves only to obtain precision of execution.
>
> **—Igor Stravinsky**

In Chapter 1, we said that fixing the total amount of a mixture facilitates modeling of the response as a function of component proportions. Mathematically this translates to an equality constraint, that is, the sum of all ingredients comes to a fixed total. Now we detail how to contend with further constraints on individual ingredients.

Setting Minimum Constraints

The fewest complications come from setting minimum constraints only. For example, let's say that after seeing our sidebar on Cornell's famous Harvey Wallbanger's experiment you ask a bartender to mix one up for you. The three ingredients are orange juice, vodka and vanilla Galliano (a sweet Italian liqueur). This mixed drink is traditionally served in a 10-fluid ounce (fl. oz.) highball glass in proportions of 6, 3 and 1; respectively, by the specification of the International Bartenders Association (IBA). However, suppose that the bartender runs low on the alcoholic ingredients and, furthermore, you appear to be someone who doesn't drink heavily. That leads to him establishing these minimums:

1. Orange juice (OJ), ≥6 fl. oz. (the more the better)
2. Vodka, ≥1.5 fl. oz. (a "jigger")
3. Galliano, ≥0.5 fl. oz. (a "splash")

The bartender starts by pouring at least 6 ounces of orange juice into the highball glass. This constraint is labeled "1" (segment 1–1) in the ternary diagram (Figure 4.1). The "mixologist" follows with a jigger of vodka—constraint 2 (segment 2–2)—and garnishes the drink with a splash of Galliano—constraint 3 (segment 3–3). This shrinks the original simplex region to the area labeled "A-B-C." The point pins down the IBA's ideal recipe from which the bartender allows himself some variation—providing a triangular zone of acceptable Harvey Wallbanger's drink mixes.

Notice that the addition of lower constraints does not affect the shape of the mixture space: It remains a simplex region. However, the shrinkage makes it very inconvenient to work graphically within this space. We need a way to blow it back up—like a photo enlarger. Fortunately, experts in mixture design developed mathematical tools to accomplish this.

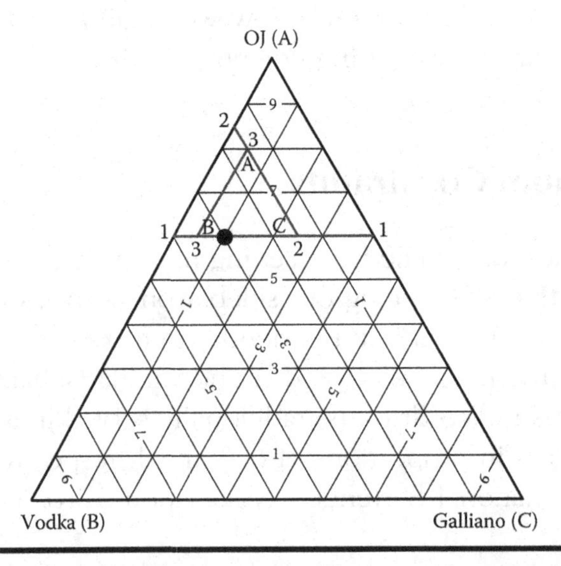

Figure 4.1 Constrained region for Harvey Wallbanger mixed drink.

GOING REALLY RETRO: HARVEY WALLBANGER BUNDT CAKE

A 1966 baking contest sponsored by Pillsbury, a food company based in Minneapolis, Minnesota, featured a "Tunnel of Fudge" cake baked in a circular "Bundt" pan invented locally by H. David Dalquist, founder of Nordic Ware. This creative combination of fudge-cake recipe (provided at the Pillsbury website) and nonorthogonal geometry created a sensation—surpassed only when someone came up with the idea of a Harvey Wallbanger bundt cake. This makes a great dessert for a retro 1970s party—it goes well with leisure suits, hot pants, and disco music.

> If you look over the years, the styles have changed—the clothes, the hair, the production, the approach to the songs. The icing to the cake has changed flavors. But if you really looked at the cake itself, it's really the same.
>
> **—John Oates**

Expanding the Constrained Space via Mathematical Coding

This might be a good time for you to take a break and prepare yourself for a mathematical sprint by drinking a suitable beverage—caffeinated rather than alcoholic (tempting as that may be!). The objective of this number-smithing will be to rescale the mixture space for the sake of graphical convenience. This will be far easier to explain by an example. Let's use the one detailed in Problem 3.1, which showed how to optimize a detergent formulation with the aid of computer software for mixture design and analysis.

We will presume you've worked through the tutorial on this case. What's most relevant at the moment is how the experimental focus remains on only the three key components which comprise 9% by weight of the total mixture. Thus, it becomes convenient to rescale or "code" the actual component levels to the actively varied ingredients—designated as "real" values (x′):

$$x_i' = \frac{X_i}{\sum X_i} = \frac{X_i}{9}$$

where "i" represents an individual component. The coding from actual to real levels is very simple—it's just a conversion to fractions, in this case

Table 4.1 Coding of Actual to Real Values

	Actual		*Real*	
	L_i	U_i	l'_i	u'_i
1. Water	3%	5%	0.333	0.556
2. Alcohol	2%	4%	0.222	0.444
3. Urea	2%	4%	0.222	0.444

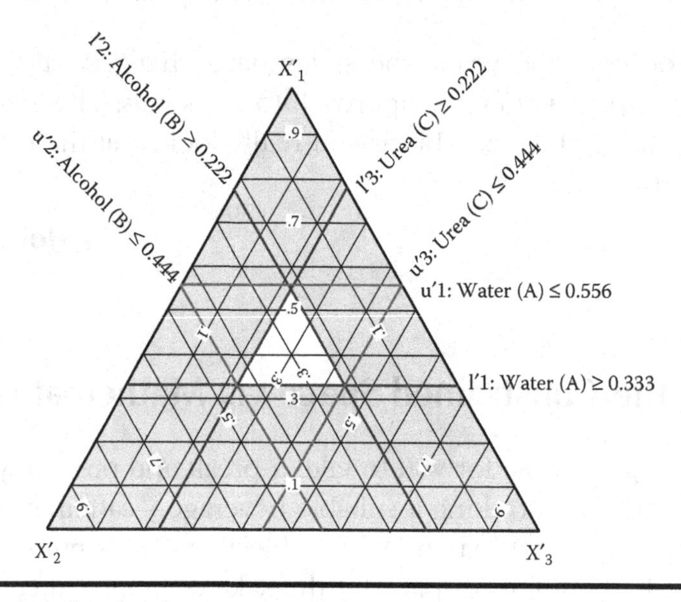

Figure 4.2 Real region for detergent experiment.

based on 9% as the formulator's total. Table 4.1 shows the ranges, lower (L and l′) to upper (U and u′), for the three detergent ingredients in actual percents (by weight) and their real values scaled 0 to 1 (dimensionless).

Figure 4.2 maps out the real boundaries of a ternary mixture diagram.

Notice how the space, a simplex triangle, is defined by the lower (l′) limits of the three components. However, it will be hard to do design-work within such a small region and, ultimately, visualize the modeled response. We must perform one more mathematical transformation, called "pseudo" coding, to expand the restricted reals into a maximum range from 0 to 1:

$$\text{L\_Pseudo} = \frac{\text{Real} - l'_i}{1 - \sum l'_i}$$

$$x_i = \frac{x_i - l'_i}{1 - \sum l'}$$

$$x_1 = \frac{x'_1 - 0.333}{1 - 0.778}$$

$$x_2 = \frac{x'_2 - 0.222}{1 - 0.778}$$

$$x_3 = \frac{x'_3 - 0.222}{1 - 0.778}$$

The prefix "L" indicates that this coding has been done on the basis of the lower boundaries (for more explanation, see the appendix on upper-bounded pseudos). We've removed the prime (′) mark to differentiate the pseudos from the reals. Table 4.2 provides the amazing results of this transformation: The components now range from 0 to 1!

The payoff to pseudo coding can be seen in Figure 4.3—the entire ternary graph is now utilized to map out the design points for the detergent experiment.

You can see in Figure 4.3 that we've enlarged the space via this coding. It now becomes far easier to work with for design purposes. Notice that the points are now shown—two each (designated by the number "2") at the three corners of the triangular experimental space, for example. (The overall centroid is also replicated.) Refer to Problem 3.1 and the associated online tutorial for more details on this design.

Table 4.2 Coding of Real to Pseudo Values

	Real		*Pseudo*	
	l'_i	u'_i	l_i	u_i
1. Water	0.333	0.556	0	1
2. Alcohol	0.222	0.444	0	1
3. Urea	0.222	0.444	0	1

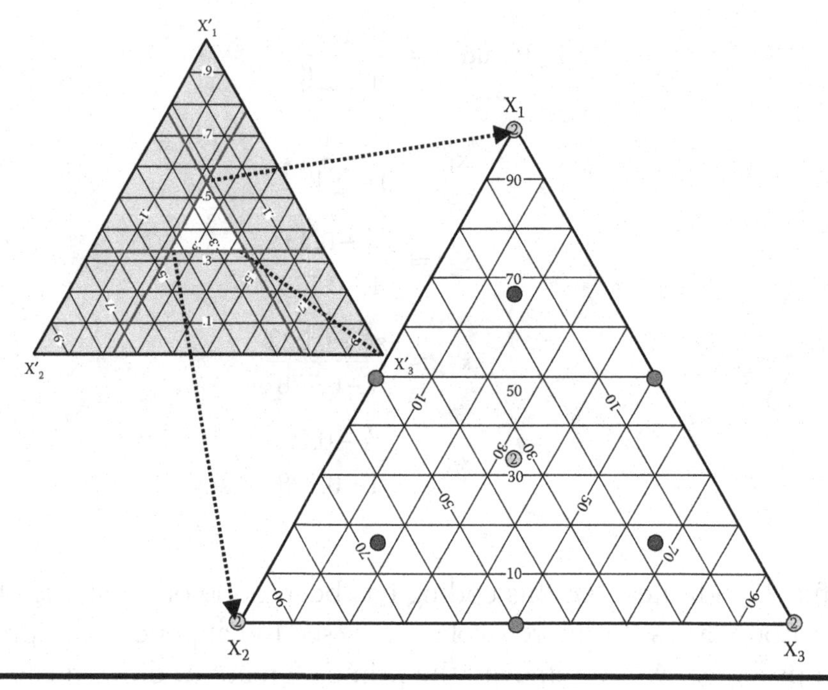

Figure 4.3 L-pseudo-coded region for detergent experiment.

Note this graph is ruled by percent—the standard for an "off-the-shelf" ternary graph. Don't let this throw you off. When looking at the lines labeled 10, 20, 30… (e.g., from bottom to top), read these off as 0.1, 0.2, 0.3… in coded level; respectively.

HOW TO EXPLAIN THE RELATIVE COMPONENT LEVELS DISPLAYED BY CODED VALUES

If a colleague or superior asks you to explain a graph such as that of Figure 4.3, be sure to remember that the 0 coded level may not mean that none of the components is present in the blend. For example, our detergent experiment always has some of each ingredient. One way to present this is to say that as one moves from any side of the ternary graph in L-pseudo-component coding, an ingredient goes from its "leanest" (lowest) level to its "richest" (highest) level. Consider, for example, the urea that's varied experimentally for the detergent. It goes from an actual low of 2% by weight to a maximum of 4%. However, it may suffice to say that the left side of the triangle in Figure 4.3 represents mixtures with urea being leanest and the lower right corner (labeled X_3) is the point where this chemical is richest.

Why It Was Worth Reading This Chapter and What's in It for You as a Formulator

Although we tried to keep this chapter short and sweet, you may be wondering what's really in it for you. First, consider that as a practical matter, it may be relatively uncommon to experiment on components that can range from none to all, that is, zero to one-hundred percent (0%–100%) of the mixture. For example, a coatings chemist would not dream of making up a can of pure pigment. Or, how about this—a pound cake comprised only of flour—no butter, sugar or eggs. Thus it becomes mandatory to impose constraints on individual ingredients.

This provides a practical perspective on the first part of this chapter, which details an example of a constrained mixture design. The remainder seemingly devolves into mathematical detail on component coding. Other than statisticians, who needs that? In fact, computer software that handles the design and analysis of mixtures may shield you from anything other than the actual levels of components, at least in the recipe sheet (ideally in random order, of course) and the outputs that you will present to a nonstatistical audience. However, if you make the effort to see how coding greatly simplifies laying out designs via expansion of the experimental region, this will at least provide an appreciation for how statisticians do the calculations (or program software to do so). Furthermore, when you see some of the model details reported, such as "Final Equation in Terms of L-Pseudo Components," the prior exposure to this standard mathematical methodology for mixtures may prevent the perfectly normal adverse reaction to statistical mumbo-jumbo. We hope so!

A TRICK TO SPELLING "PSEUDO"

Mark unabashedly declares his coauthors Pat and Martin to be superior "mathletes." However, Mark takes great pride in his word skills, especially spelling. Thus it was embarrassing for him to continually misspell "pseudo" as "psuedo" (sic). Although he maintains a standing protest that it is unfair to put a "p" before an "s" and combine the two vowels of "e" and "u," Mark had to come up with a way to remember the correct spelling of the essential statistical term. Finally, he realized that the only way was to first spell it in his head with the "e" and "u" transposed as "sued"—that being related to onerous legal matters is bad. Mark then

(*Continued*)

silently hums the famous Elvis Presley song about his blue suede shoes and knows that this is all wrong as well. That puts him back on track for "pseudo"—the correct spelling. Such are the lengths one must go through to contend with peculiar scientific jargon.

> Drink my liquor from an old fruit jar. ... You can do anything but lay off of my blue suede shoes.

—Excerpt from lyrics by Carl Perkins (music also originated by him—recorded in 1956)

Practice Problem

To practice using the statistical techniques you learned in Chapter 4, work through the following problem.

Problem 4.1

Two Portuguese food scientists, Margarida Vieira and Cristina Silva (2004), applied mixture design to optimize the taste of an exotic nectar based on the Cupuacu (pronounced "koo-poo-a-soo") fruit from the Amazon jungle. They established the following ranges (coded on a 0 to 1 scale) for their mixture components (in percentages by weight):

1. (X_1) Sugar, 10%–25%
2. (X_2) Cupuacu, 15%–30%
3. (X_3) Water, 60%–75%

The two fruit-juice formulators then set up a completely replicated second-degree simplex-lattice design. They augmented their design by adding three axial check blends (unreplicated), and the overall centroid replicated three times.

Pencil in on Figure 4.4, a blank ternary diagram, with the upper and lower constraints for each of the three components in their actual levels, expressed as weight percent. You should find that this space remains a simplex, albeit much smaller than the original. In the answer posted for this problem, we provide a link to an article (Anderson, 2005) that details the computer-aided design and analysis of this experiment, including the final formulation of the Cupuacu nectar that matched what the original experimenters recommended as the most desirable sensorially.

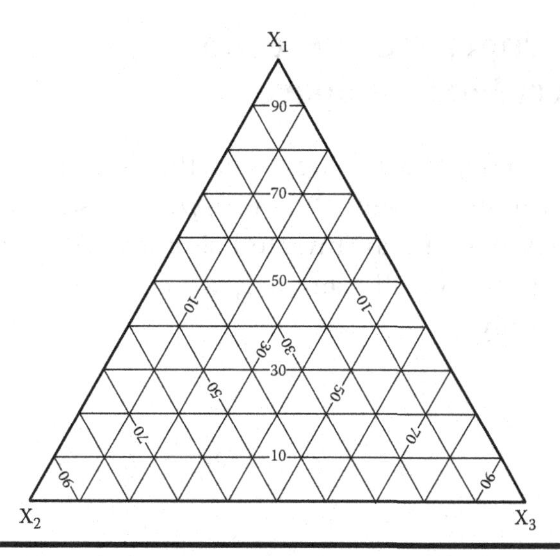

Figure 4.4 Blank ternary diagram to draw in Cupuacu component constraints.

A SEVEN-COMPONENT (?) BEVERAGE: "7-UP®"

An early sales bulletin (~1929) for 7-Up told consumers that the drink contained seven natural flavors "so blended and in such proportions that, when bottled, it produces a big natural flavor with a real taste that makes people remember it." The original name was Bib-Label Lithiated Lemon-Lime Soda. The new moniker coined by advertising expert C. L. Grigg, "7-Up," made the drink seem more palatable.

From the story, above, one would conclude that the "7" in "7-Up" relates to the number of ingredients. However, per the website Wikipedia, Donald R. Sadoway, a professor of materials chemistry at the Massachusetts Institute of Technology, contends that the name derives from the atomic mass of lithium, 7, which was originally one of the key ingredients of the drink (as lithium citrate—purported to be a mood-stabilizing drug). In any case, evidently during one of the many reformulations of 7-Up in 1950, lithium citrate was removed.

Here's a thought for further experimentation by the Portuguese food scientists—punch up the Cupuacu blend with some bubbly such as 7-Up.

(The rights to the 7-Up brand are held by the Dr Pepper Snapple Group in the United States, and PepsiCo (or its licensees) in the rest of the world.)

Appendix 4A: Upper ("U") Pseudo Coding to Invert Mixture Space

At the outset of this chapter, we suggested that setting lower constraints keeps things simplest. When *upper* constraints define the design space, it may become an inverted simplex. For example, let's say that, while keeping the overall total at 1, we constrain three-components as follows—upper only:

1. $X_1 \leq 0.4$
2. $X_2 \leq 0.6$
3. $X_3 \leq 0.3$

As Figure 4A.1 illustrates, this upper-only approach to setting component constraints creates an upside-down triangle within the real-scaled ternary diagram! Unfortunately, as you can see in Figure 4A.2, the lower-bounded pseudo values for an inverted simplex do not vary from 0 to 1, which would be most advantageous graphically and mathematically. Rather, they range from 0–0.5 in this coded scale.

This geometry can be likened to trying to put a square peg in a round hole. Luckily for all of us formulators, an expert in the field (Crosier, 1984)

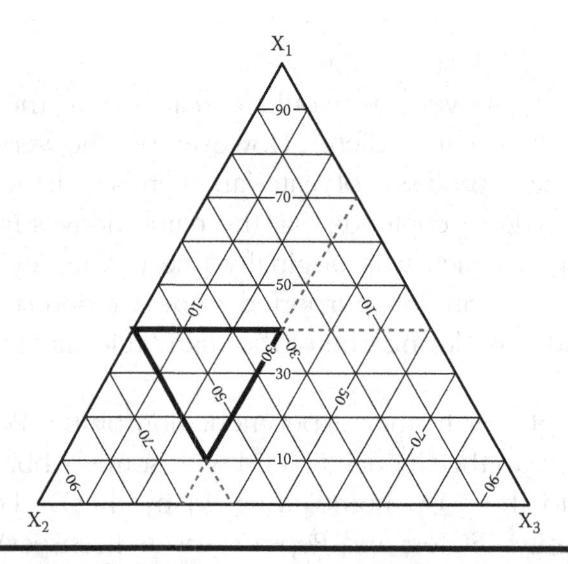

Figure 4A.1 Inverted simplex (real) created by entering only upper component constraints.

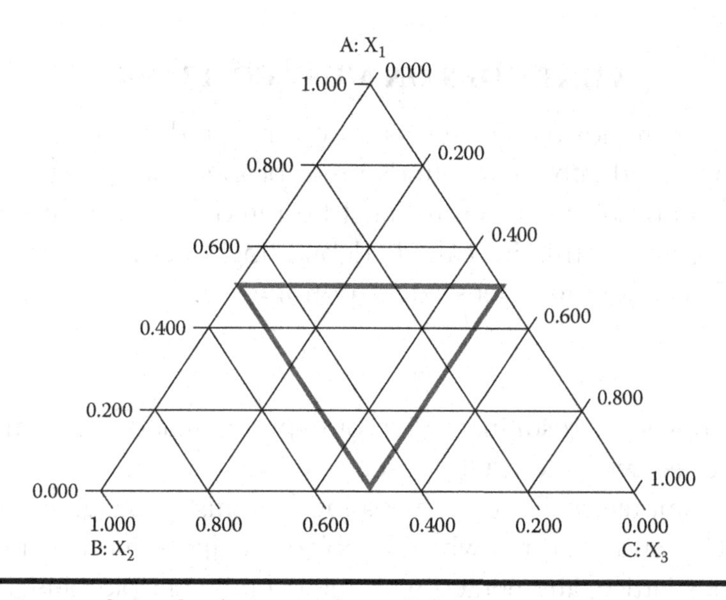

Figure 4A.2 Inverted simplex in L-pseudo-coded mixture space.

came up with a simple solution: Code pseudo-components from the upper, not the lower, constraint. The key equation is:

$$\text{U\_Pseudo} = \frac{u_i' - \text{Real}}{\sum u_i' - 1}$$

If you are into math, compare this to the equation for L-pseudo coding we provided in the core of the chapter. Software designed for mixture experimentation, such as the program provided with this book, can detect when upper ("U") coding will be advantageous. Then, if the user elects to go this route, it can do the necessary calculations.

Let's see how U-pseudo coding works on a real-life example. A coatings chemist designed a mixture experiment on a chemical paint remover for an aerospace application (Hensley, 2008). The chemical supplier's material safety data sheet (MSDS) specified the following constraints for the three key active ingredients for their recommended formulation, which totaled 12% by weight:

1. 0%–5%
2. 0%–5%
3. 2%–7%

WHAT GOES ON MUST COME OFF

The coatings removal process is often referred to in the field as "depainting." Evidently, this sounds more proper than paint "stripping." Similarly, when an old structure must be removed before putting up a new facility, would this be called "debuilding?" Perhaps this terminology could become the razer's edge (pun intended).

Figure 4A.3 shows the resulting region of experimentation for a mixture design developed with the usual L-pseudo coating.

Aided by computer software the coatings chemist redesigned his experiment using U-pseudo coding, which enabled the application of a second-degree simplex-lattice, augmented with the overall centroid plus three check blends—illustrated in Figure 4A.4. Notice that the vertices and centroid are replicated (designated by the number "2").

This "flip" from lower to upper coding doubled the ranges—they had been 0 to 0.5 in L-pseudo, but now (after converting to U-pseudo) all the components go from 0 to 1. Statisticians delving into an evaluation of this design on a "before and after" basis will be most impressed by the decreased collinearity in the matrix and among the model coefficients

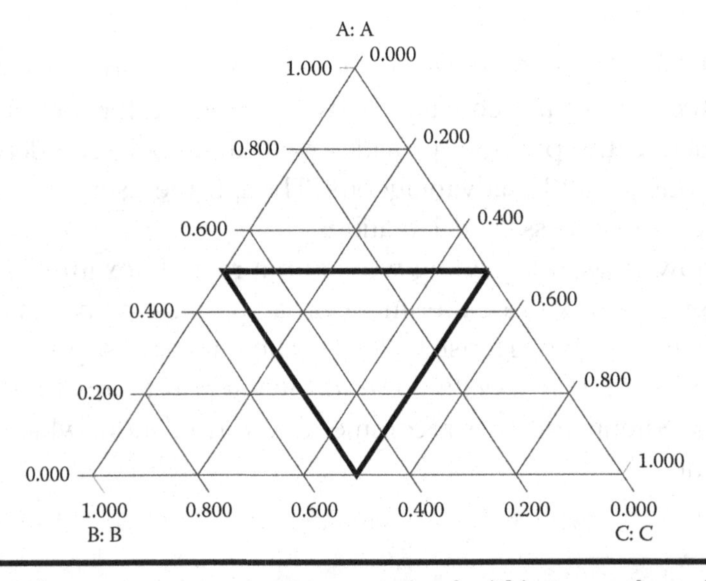

Figure 4A.3 Paint remover experiment constrained within L-pseudo-coded mixture space.

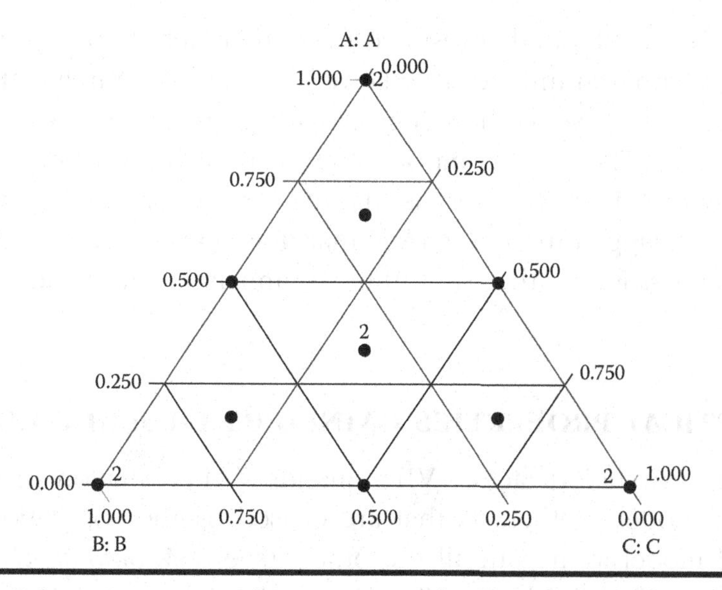

Figure 4A.4 Paint-remover experiment converted to U-pseudo-coded mixture space.

(see sidebar for details). However, the payoff will be far more visible via the larger region of experimentation displayed in the response graphics.

Beware when working in this upside-down world of U-pseudo coding: Low becomes high, and high becomes low on the ternary diagrams. For example, look carefully at the triangle in Figure 4A.5, which displays the

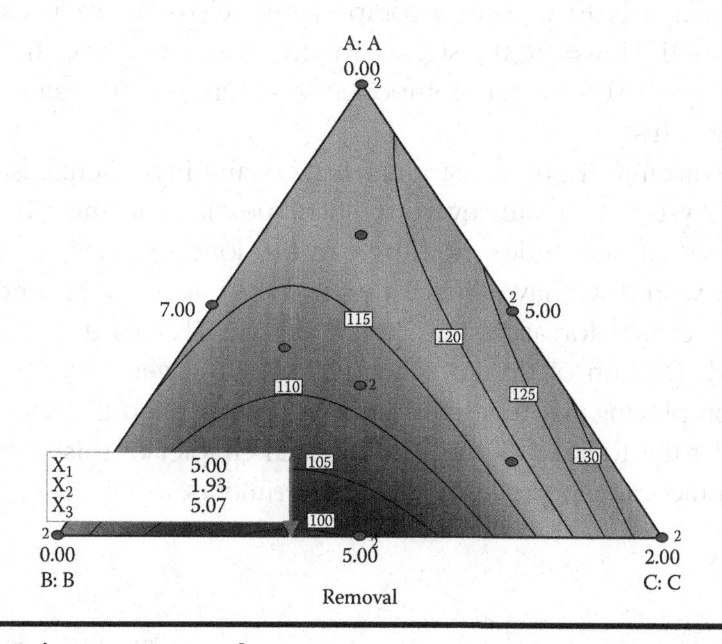

Figure 4A.5 Paint-remover results.

response surface contours that resulted from the paint-remover experiment. (They predict a time in minutes to take off the coating entirely—the lower the value, the better.) Notice how ingredient A goes from 0% at the top to 5% at the bottom. The other component axes are also reversed.

The happy ending to this story is that the formulator came up with an optimal recipe (flagged in Figure 4A.5) that dramatically reduced the processing time for stripping the paint, thus minimizing hazards due to chemical exposures.

STATISTICAL PROPERTIES GAINED BY U-PSEUDO CODING

The variance inflation factor (VIF) quantifies the inflation in error for estimating model coefficients that are caused by the correlation of any individual model term with all the others (see *RSM Simplified, 2nd ed.*, Chapter 2, p. 33–34). Ideally, all VIFs equal 1, that is, there is no inflation of variance whatsoever. When this occurs, the experimental design matrix is said to be orthogonal, which provides maximum power for estimating each coefficient. Unfortunately, due to the imposition of a fixed total of the ingredients, mixture designs can never be orthogonal. However, VIFs can still be useful for comparative purposes.

For example, let's look at the paint remover case. In actual units, the VIFs range from 55 to 162 per a computer-aided design evaluation. Coding the matrix to L-pseudo-components helps–this reduces the VIF range to 54–91. However, the significant improvement comes from recoding to the upper (U) pseudo-components—a matrix with values of 1.55–1.72 for the VIFs.

A forewarning: Tight constraints on certain ingredients, such as a potent catalyst, will create severe multicollinearity, causing VIFs to soar. Due to physical necessities, nothing can be done to alleviate this. Even so, do not worry; we have an ace up our sleeve to work around the difficulty this causes for assessing design power. This card is a graphical tool called "fraction of design space" (FDS). However, it will be best to hold off on playing out this hand until we provide a full deck. This will come under the topic of "sizing" in the final chapter of this book, which provides practical aspects of mixture experiments.

Chapter 5

Optimal Design to Customize Your Experiment

> The computer can't tell you the emotional story. It can give you the exact mathematical design, but what's missing is the eyebrows.
>
> **—Frank Zappa**

In Chapter 4, we detailed a set of constraints to keep individual ingredients within reasonable limits. For example, the recipe for the Harvey Wallbanger's cocktail requires at least 6 ounces of orange juice in the 10-ounce highball, thus limiting the alcoholic content coming from the vodka and Galliano. Fortuitously, after accounting for constraints on these other two ingredients, the resulting experimental region remains a simplex, thus accommodating standard mixture designs. In most cases, though, the limits imposed by formulators create a nonsimplex shape such as the one in Figure 5.1, where ingredient A is cut off at 60%.

Dealing with nonsimplex constraints is our focus for applying optimal designs. However, these designs, typically computer-generated, can be customized to any shape for any order of polynomial desired by the experiment. Thus, for example, with the proper software, you could build a special cubic "custom design" within a simplex geometry. Nevertheless, to keep things more interesting, we will restrict our discussion in this chapter to optimal design within nonsimplex regions.

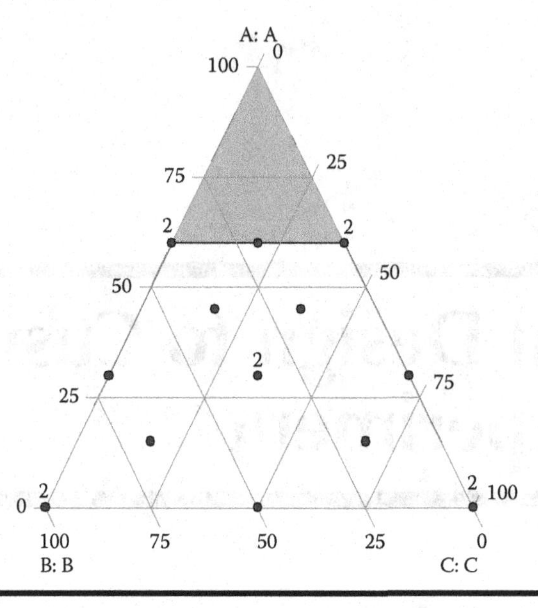

Figure 5.1 Nonsimplex created by truncating component A.

THE TRAPPINGS OF OUR TRAPEZOID

The truncated triangle in Figure 5.1 forms a trapezoid or trapezium, depending on whether you go by American or British terminology, respectively. We decorated it with 18 points to provoke thought on how a mixture design might be fitted to this nonsimplex shape. It pays first to cover the corners, or vertices, in this case with 2 points apiece to provide estimates of pure error. We also replicated a blend at the middle of the experimental region—the centroid. The next step is considering the mid-points of the edges. As a final touch, we put in points midway between the centroid and the vertices—these serve as check blends to fill out the space. This handmade layout of experimental formulations is simple, but it literally (by the definition below) provides a table upon which to build this chapter on optimal design. Things will get more complicated from here on, particularly beyond three-component designs.

From Late Greek trapezoeides, noun use by Euclid of Greek trape-zoeides 'trapezium-shaped,' from trapeza, literally "table."

—trapezoid (n.)—Online Etymology Dictionary
www.etymonline.com

Extreme Vertices Design: Shampoo Experiment

Now we will explore data from a case with complex constraints, that is, nonsimplex shaped. This design was created in the early 1980s by one of the coauthors (Pat) without the aid of computer software. You will get to share the pain of working through, like him, by hand—all in the interest of education, of course.

Cosmetic chemists needed help for an experiment to optimize foam height in a shampoo. They identified three key surfactants and specified constraints (in percentages by weight) on each of them, as follows:

1. Triethanolamine (TEA)–Lauryl Sulfate: 20%–30%
2. Cocamide DEA (diethanolamine): 1%–7%.
3. Lauramide DEA: 1%–3%.

They required that the sum of these three components come to a constant of 30% of the shampoo. In other words, 70% of the formulation (water, thickeners, preservatives, etc.) remained fixed. The chemists focused only on one response, which is the foam height—measured in millimeters (mm).

WASH THAT FOAM RIGHT OUT OF YOUR HAIR

The ultimate test on foam is the "half-head test" because it provides a real-live one-to-one comparison of a newly developed shampoo versus a reference product. A representative subject applies the two products on either side of their head. They then evaluate on a sensory scale (typically 1–9) the feel of the foam, how it looks and what it takes to rinse off the shampoo, all under the watchful eye of the cosmetic chemists.

Source: Ansmann, A. et al., Personal care formulations. Chapter 7 of the *Handbook of Detergents: Part D: Formulation*, M. S. Showell (Ed.). CRC Press, 2005, p. 248. This book also provides an excellent overview of "Statistical mixture design for optimization of detergent formulations," in Chapter 2 by S. S. Ashrawi and G. A. Smith.

When the ranges of the factors are not equal such as in this case with low to high differences of 10%, 6%, and 2% for A, B, and C, respectively, the design space will not be a simplex. It turns out that the constraints for the shampoo experiment create a four-cornered shape defined by the vertices laid out in Table 5.1, which we identified in actual component levels via the algorithm

Table 5.1 Vertices for Shampoo Experiment

Vertex	*Actuals (%)*			*Reals (Fraction of Total)*		
#	$A(X_1)$	$B(X_2)$	$C(X_3)$	x_1	x_2	x_3
1	28	1	1	0.933	0.033	0.033
2	22	7	1	0.733	0.233	0.033
3	20	7	3	0.667	0.233	0.100
4	26	1	3	0.867	0.033	0.100

detailed in Appendix 5A. This table also provides the compositions as a fraction of the total (30%)—then referred to a "reals."

When scaled to reals, the vertices can be plotted on ternary paper so the experimental region can be laid out and additional points around and inside it plotted. Figure 5.2 provides the picture of the plot.

This experiment design looks a bit cramped in the rectangular "penthouse" of the triangular structure. It needs to be converted to pseudo-component coding, which we introduced in Chapter 4. See Figure 5.3 for the resulting expansion of space on the ternary plot and blends filled in along the edges (the four midpoints) and the interior (overall centroid and four axial checkpoints). For added precision in fitting, all the blends were replicated in the final randomized plan.

The result of this experiment, for what it's worth (a lot to the formulators, but not for our discussion on optimal design), is displayed in Figure 5.4,

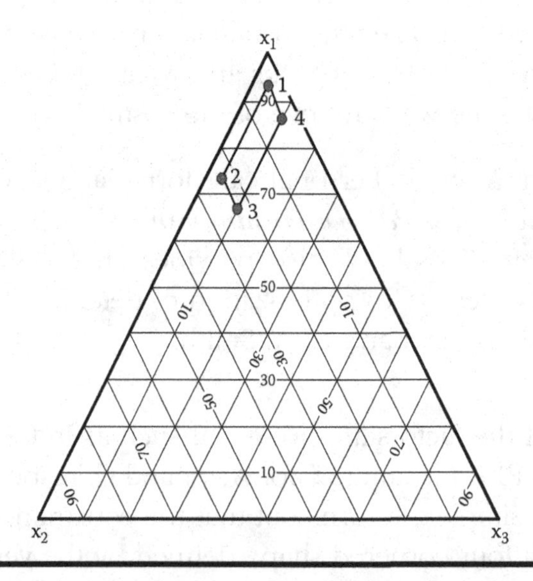

Figure 5.2 Extreme vertices for shampoo formulation plotted in reals.

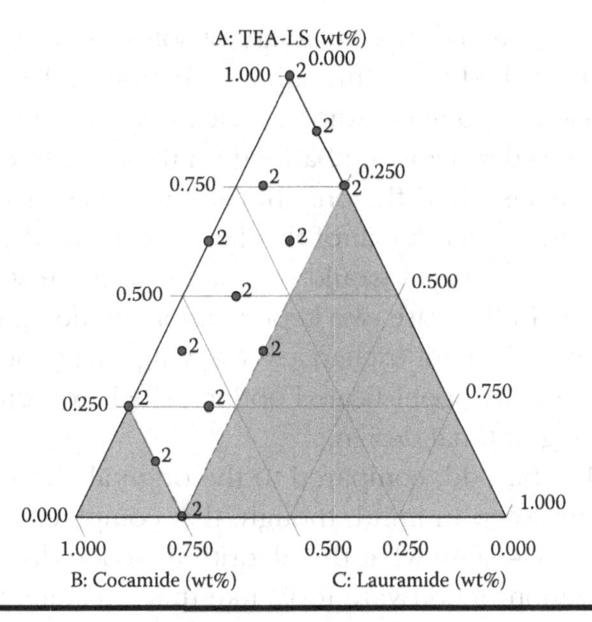

Figure 5.3 Shampoo design in pseudo-component space.

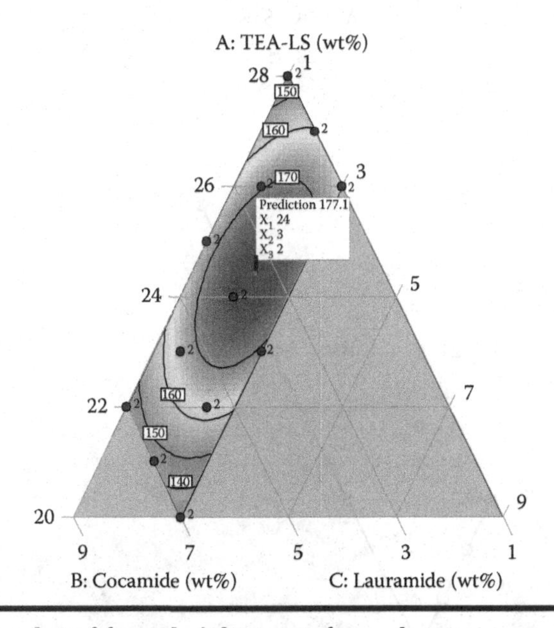

Figure 5.4 Contour plot of foam height (mm) from shampoo experiment.

with the axis labeling and gridlines reverted conveniently to actual scale. The optimum foam height in mm is flagged and identified with the compositional coordinates.

This case study on the shampoo experiment provides a good start on how to contend with constraints that do not form a simplex. Keep in mind that, this being done before the availability of software for mixture DOE, the cosmetic chemists had to go through all the work to lay out the design graphically, decode the point coordinates back to actual component levels, and randomize the order before embarking on the experiment. Nowadays specialized programs, such as the one that accompanies this book, provide optimal custom designs that do all of this for formulators. Figure 5.5 presents the shampoo design the way it would be done with these state-of-the-art computerized tools. In this case, we kept it simple by doing the building via an algorithm called "point exchange." We will provide details on its construction, and a more-sophisticated option called "coordinate exchange," later—after covering optimal design.

This may look a bit odd, compared to the original handmade experiment design. Keep in mind, though, that computers care nothing for what pleases the eye—only what the algorithmic code drives them to do. Without the utilization of software tools like this, it is daunting to design

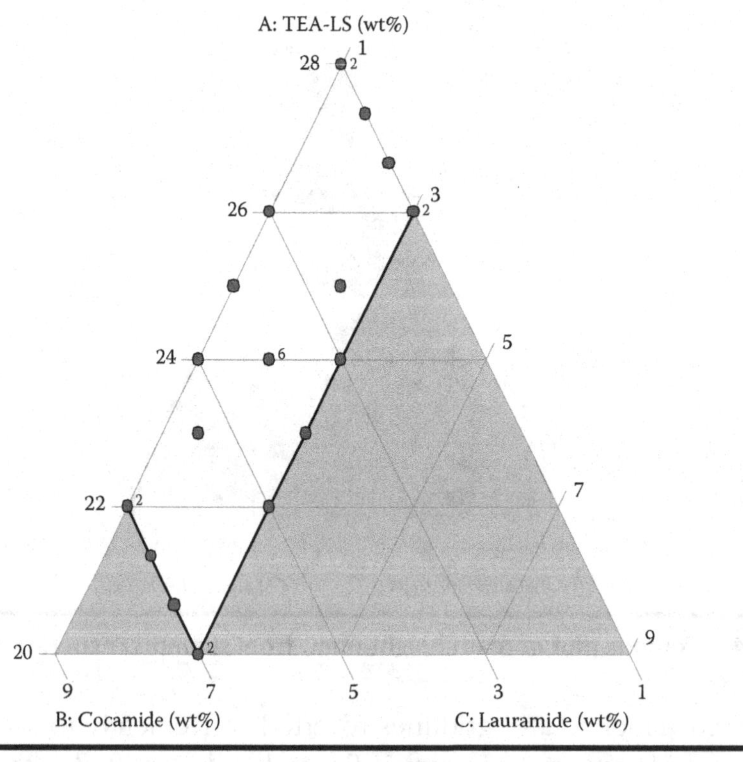

Figure 5.5 Shampoo experiment rebuilt using optimal custom-design tools in computer software.

mixture experiments with three components, unreasonable to embark upon with four, and silly to even contemplate beyond that. Let's now turn our attention to what makes a good custom design and the generally accepted statistical criteria for optimality.

Optimal Designs Customize Your Experiment as You See Fit for Any Feasible Region

We advise using ASL designs, spelled out in Chapter 3, for all mixture experiments involving 5 or fewer components that form a simplex as-is or inverted (detailed by Appendix 4A). If you want to study 6 or more components, we suggest that, before embarking on optimization, you consider first doing a screening experiment using designs that will be provided later in the book. For all other experiments (i.e., nonsimplex and/or not models of degree 1, 2, or 3), use a statistically optimal design—in particular, a specific type called "I-optimal" that is best suited for formulation improvement.

IF YOU WANT TO LEARN THE ALPHABET OF OPTIMALITY CRITERIA

Since the beginnings of optimal design half a century ago by Box and Lucas, (1959) industrial statisticians have developed many matrix-based criteria which they label alphabetically. For DOE and RSM the favored few are D and I, respectively. Since mixture design is mainly used to optimize formulations, we will use I-optimality exclusively. This criterion minimizes the average prediction error of the design across the region of experimentation—a very suitable goal for mapping out a formulation space.

For details on D versus I refer to "Practical Aspects for Designing Statistically Optimal Experiments" (Anderson and Whitcomb, 2014). Also, see Chapter 4 of *RSM Simplified, 2nd ed.*, for in-depth coverage on these computer-generated run-layouts. There you are provided with details on "Optimality from A to Z," a checklist on what these design-building

(Continued)

criteria aim to do, and helpful hints on Getting the Computer to be "More Open-Minded About What's Optimal."

Our mission now is to provide formulators like you just enough information (i.e., a "need-to-know" basis) to feel comfortable using optimal design. You are likely to discover that your constraints will not create a simplex. Thus these custom design tools will be a godsend. Just do not get hung up on the details of their development and mathematical construction.

> No postulated model can ever be assumed to be fully correct [therefore], regarding the basic assumptions underlying the alphabetic-optimality approach, which are often unrealistic from the practical viewpoint of actually designing real experiments.
>
> **—Draper and Guttman (1988)**

Take This Tableting Case-Study Now and Call the Statistician in the Morning

Ever since Bayer invented aspirin in 1897, medical doctors have been telling their patients to "take two and call me in the morning." We hope to keep you from getting a headache by detailing optimal design for mixture experiments via a case study, which, as it happens, involves the formulation of a tablet.

This case study on optimal design is loosely based on an example by Lewis et al. (Lewis, G.A., D. Mathieu, and R. Pahn-Tan-Luu, *Pharmaceutical Experiment Design*, Chapter 10, "Mixtures in a constrained region of interest," Section II, "Second example: A hydrophilic matrix tablet of 4 variable components," 2008, Informa Healthcare, USA, Inc.). The formulators hoped to achieve a sustained release of an active pharmaceutical ingredient (API) tablet. The key ingredient is a hydrophilic polymer that swells in the presence of water and, thus, impedes drug release. By experimenting on the proportion of this polymer to three excipients—inactive substances that served as the drug vehicle—they developed a handle on the rate of API release, as well as vital physical properties of the tablet.

Ranges regarding weight percent are laid out here for the four components to be varied:

1. Lactose ("Lact"), 5%–42%
2. Calcium phosphate ("CaPh"), 5%–47%
3. Microcrystalline cellulose ("Cell"), 5%–52%
4. Hydrophilic polymer ("Poly"), 17%–25%

These ingredients were kept to a total of 98% to accommodate a fixed concentration of the API ("Drug") at 2%.

The first step for designing a mixture experiment is to define the experimental region by running the ranges through an extreme-vertices algorithm, such as the one embedded in the software accompanying this book. We advise entering them in the simplex-lattice design-builder, which, in this case, recognizes that the mixture space does not form a simplex. The program then suggests an optimal design for customizing the experiment to the nonsimplex space.

Optimal Design Simplified

With the aid of software (plus being able to slide by the mathematical and computational details), it is quite easy to generate an optimal experiment design. All you need to do is specify the model that the design must be capable of fitting. For the most part you will do well by choosing a quadratic—a model that characterizes nonlinear blending, that is, antagonism or synergism between components. Your chosen model then dictates how many unique blends will be required in the mixture design—for the most part, one for each term that must be estimated.

In this case, the tablet formulators went with this standard selection of a quadratic model, which featured 10 terms—4 for the main component effects (A, B, C, and D) plus 6 for nonlinear blending (AB, AC, AD, BC, BD, and CD). To fit these 10 terms, 10 unique blends are required. This is where the I-optimal criterion comes into play. It will be the judge by which blends to run from within the mixture is spaced. Although this can be done with few geometric restrictions via a method called "coordinate exchange" (detailed in sidebar "Candidate-free Approaches for Exact Optimal Designs," Appendix 7C, *RSM Simplified, 2nd ed.*), let's keep things simple by identifying a discrete number of points as candidates for the optimal selection. This is called the "point exchange" method. The only requirement is that the candidate set exceeds the number of blends required to fit the model.

The starting point for building up a good candidate set is the extreme vertices. These being far apart will provide the highest leverage for fitting the main component effects. There are 12 extreme vertices in the mixture space for the tablet experiment. To fit the nonlinear terms, it is helpful to also include centers of edges (binary blends)—18 in this case. One more point to include is the overall centroid—a complete blend. This brings the total to 31. Furthermore, to fill in the gaps between the centroid and vertices, let's add the 12 axial check blends. This brings the total to 43 candidate points—well beyond the 10 needed for the quadratic mixture model.

From this set of candidate points, we laid out (with the aid of the software) the 20-run custom design shown in Table 5.2. (Note that, to make it simpler to see that the components add up to 100%, we included the API (shaded to highlight it being fixed) and expressed the fractions). It is comprised of the following blends:

- 10 for the model chosen I-optimally.
- 5 to test lack of fit ("LOF") selected by a distance-based algorithm (refer to the sidebar "Getting the Computer to be More Open-minded About What's Optimal" in Chapter 7 of *RSM Simplified, 2nd ed.*) to fill the component space.
- 5 replicates ("Rep") for estimation of pure error (needed for the LOF test) chosen by the same optimal criterion as that used for selecting the model points—these being the points creating the maximum impact for improving the fit.

Table 5.2 Custom Design of Experiment on Tablet Formulation

#	*Build*	*Space*	*A: Lact* (*wt%*)	*B: CaPh* (*wt%*)	*C: Cell* (*wt%*)	*D: Poly* (*wt%*)	*E: Drug* (*wt%*)
1	Model	Vertex	42	34	5	17	2
2	Model	Vertex	5	47	29	17	2
3	Model	Vertex	24	5	52	17	2
4	Model	Vertex	42	26	5	25	2
5	Model	Vertex	5	47	21	25	2
6	Model	Vertex	16	5	52	25	2
7	Model	CentEdge	42	5	30	21	2
8	Model	CentEdge	25	47	5	21	2
9	Model	CentEdge	5	20	52	21	2
10	Model	Center	$23\frac{1}{6}$	$25\frac{2}{3}$	$28\frac{1}{6}$	21	2
11	LOF	CentEdge	42	$19\frac{1}{2}$	$19\frac{1}{2}$	17	2
12	LOF	CentEdge	5	$31\frac{1}{2}$	$36\frac{1}{2}$	25	2
13	LOF	AxialCB	$26\frac{1}{12}$	$36\frac{4}{12}$	$16\frac{7}{12}$	19	2
14	LOF	AxialCB	$14\frac{1}{12}$	$36\frac{4}{12}$	$24\frac{7}{12}$	23	2
15	LOF	AxialCB	$19\frac{7}{12}$	$15\frac{4}{12}$	$40\frac{1}{12}$	23	2
16	Rep	Vertex	42	26	5	25	2
17	Rep	CentEdge	42	5	30	21	2
18	Rep	CentEdge	25	47	5	21	2
19	Rep	CentEdge	5	20	52	21	2
20	Rep	Center	$23\frac{1}{6}$	$25\frac{2}{3}$	$28\frac{1}{6}$	21	2

Take-Home Advice on Deploying Optimal Design

This concludes our briefing on optimal design. To keep things simple, when you set up a mixture experiment with the aid of software that provides the proper tools to get this job done, start by setting it up as a simplex. When warned that your constraints make this impossible, shift to the program's custom builder using optimality (I-best, but D-acceptable) for selection of your model points. The point exchange option provides compositions from specific geometric locations within your experimental space, whereas coordinate exchange roams relatively freely—either selection serves well, so we suggest trying them both and picking whichever layout works best as a practical matter. Provided the minimal design (model points only) does not exhaust your budget of time, cost or material consumption, always add 3–5 extra "check" blends for LOF testing and replicate 3–5 of these and the others (picked for the model) to provide an estimate of pure error (needed for LOF).

SPECIFICATIONS FOR CUSTOM DESIGN ON SHAMPOO

Now that we've covered optimal design, here are details on how we produced the layout of points in Figure 5.5. It was set up with I-optimal criterion for a quadratic model via point exchange. To match the original design for the precision of fit, we added 8 model points beyond the 6 required to fit the model. The design was then augmented with 5 lack-of-fit points and 5 replicates. Due to random elements in optimal algorithms, you may not get an exact match to our design, but it should be more like it than the original handmade experiment.

Practice Problem

To practice using the statistical techniques you learned in Chapter 5, work through the following problem.

Problem 5.1

For a fun do-it-yourself experiment that involves optimal design for mixtures, we recommend you make play putty per the recipe posted at www.statease.com/publications/marks-play-putty-experiment (also refer to Anderson and Whitcomb, 2002). We chose the following components and levels to make up 100 milliliters (mL) of this silly stuff for our experiment:

1. Glue, 40–59 mL
2. Water, 40–59 mL
3. Borax, 1–3 mL

Enter these constraints in your mixture-design software or plot them out by hand on a ternary diagram. Do they form a simplex? Set up a design and compare it to the one in the referenced publication. Follow up by developing your own recipe for play putty. That will be fun and educational.

SILICONE SUPER BALL

A primary attribute of good play putty is a high bounce. The best we achieved was a bit over one foot in height. If you want to make a far bouncier ball, mix 5–10 mL (experiment!) with 20 mL of sodium silicate solution (water glass) and follow the instructions posted at www.flinnsci.com/silicone-super-ball/dc07585/ by a supplier of these chemicals—Flinn Scientific. A variation on the recipe (http://chemicalrecipes.blogspot.com/2010/05/making-superball.html) calls for replacing the alcohol with phenolphthalein to color the ball bright pink.

Appendix 5A: An Algorithm for Finding Vertices

Here is a relatively simple way to identify vertices in a nonsimplex mixture experiment—McLean and Anderson's extreme vertices algorithm (McLean and Anderson, 1966):

1. Make a table of all possible combinations of the lower individual (L_i) and upper individual (U_i) limits (as you would do for a two-level full factorial) for one less than the number (q) of the components (i.e., q–1). Leave the value for the remaining component (the qth one) blank. This table will consist of q^*2^{q-1} possible combinations.
2. Fill in the acceptable combinations, that is, those that don't violate individual or total constraints (TC).
3. Eliminate duplicate combinations.

For the three-components shampoo case, the first step of this algorithm dictates the construction of three two-by-two tables that spell out the four extreme combinations for A versus B, A versus C, and B versus C. These are shown from left to right, respectively, in Table 5.3.

Table 5.3 Candidate Vertices for Shampoo Experiment

A	B	C		A	B	C		A	B	C
~~20~~	~~1~~	>U_3		~~20~~	>U_2	~~1~~		28	1	1
~~30~~	~~1~~	>TC		~~30~~	>TC	~~1~~		22	7	1
20	7	3		~~20~~	~~7~~	~~3~~		26	1	3
~~30~~	~~7~~	>TC		~~30~~	>TC	~~3~~		20	~~7~~	3

Six combinations out of the twelve meet all constraints, of which four are unique (blend (20, 7, 3), which are repeated several times)—the ones boxed by thicker lines. The combinations fail either due to being outside of the individual component limits (L_i–U_i) or the total constraint (TC). For example, if you blended up 20% of A with 1% of B, then, to meet the total of 30%, 9% of C would be needed to make up the difference. However, the cosmetic chemists restricted C to a maximum of only 3%. Therefore, this combination does not satisfy the rules of the algorithm. Other combinations get thrown out straightaway due to one component taking up the entire 30% allowable total—leaving no room for anything else.

More efficient algorithms have been developed since McLean and Anderson invented theirs, most notably XVERT by Snee and Marquardt (1974) and Piepel's CONVRT algorithm (Piepel, 1988). The code is available in R to compute extreme vertices and other points in nonsimplex spaces—see Lawson and Cameron's "Mixture Experiments in R Using mixexp" (Lawson and Willden, 2016. Code Snippet 2).

Chapter 6

Getting Crafty with Multicomponent Constraints

No great discovery was ever made without a bold guess.

—Isaac Newton (attribution: p. 298, Presidential Address: The Nature of Discovery, A. Blalock—*Annals of Surgery*, 1956. www.ncbi.nlm.nih.gov/pmc/articles/PMC1465405/pdf/annsurg01270-0005.pdf)

Being bold in setting-up component levels pays off by generating more significant effects from mixture experiments. However, extreme combinations may create an infeasible product, for example, bread that fails to rise due to too little yeast at high salt levels. By imposing multicomponent constraints (MCCs), formulators can lay down boundaries that keep their region of experimentation within reason, but as broad as possible.

How Multicomponent Constraints Differ from Simple Ones

Ron Snee in a pioneering publication (Snee, 1979) presented this set of simple limits on three ingredients:

1. $0.1 \leq X_1 \leq 0.5$
2. $0.1 \leq X_2 \leq 0.7$
3. $0.0 \leq X_3 \leq 0.7$

These constraints translate to the following five boundaries on the ternary diagram shown in Figure 6.1:

1. $0.1 \leq A$
2. $A \leq 0.5$
3. $0.1 \leq B$
4. $B \leq 0.7$
5. $C \leq 0.7$

Note that the boundaries of this experimental region (in grey) are parallel to the triangular perimeter of the ternary plot or on its edge. Now comes the tricky part—two multicomponent constraints that depend on a combination of ingredients:

- $0.90 \leq 0.85A + 0.90B + 1.00C \leq 0.95$
- $0.4 \leq 0.7A + 1.0C \leq 1.0$

These additional (MCC) constraints translate to the following three additional boundaries (numbers 6, 7, and 8) displayed in Figure 6.2.

6. $0.90 \leq 0.85A + 0.90B + 1.00C$
7. $0.85A + 0.90B + 1.00C \leq 0.95$
8. $0.4 \leq 0.7A + 1.0C$

These new boundaries are not parallel to the sides of the ternary diagram. Thus, they allow for a more conservative experimental region that is crafted by the subject matter expert, in this case, a chemist.

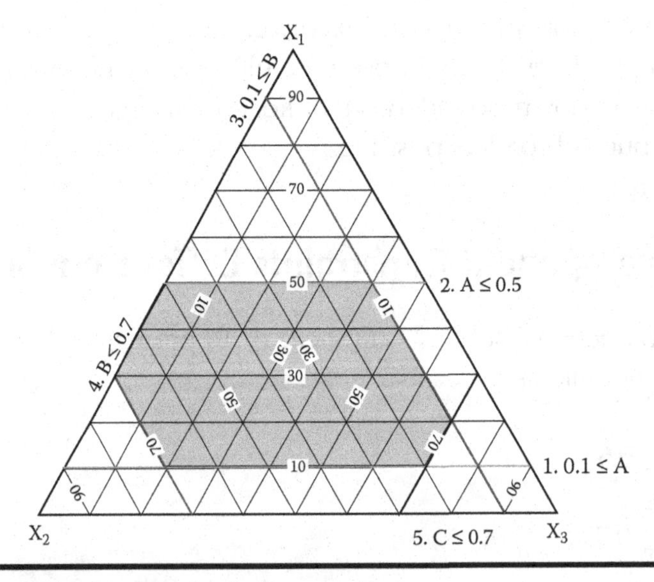

Figure 6.1 Simple constraints for Snee example.

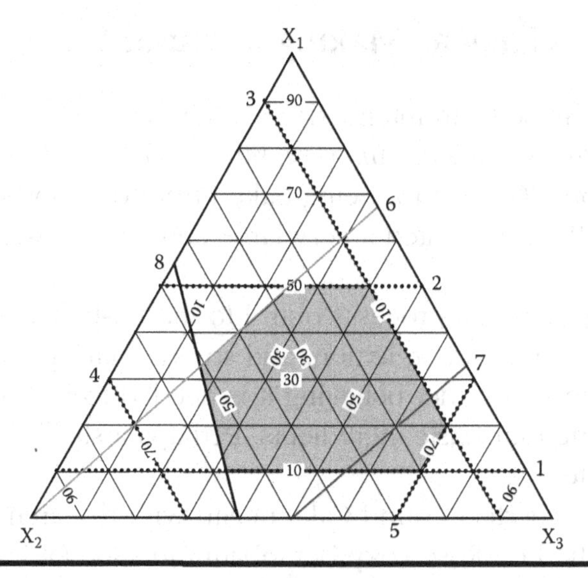

Figure 6.2 Multicomponent constraints added for Snee example.

HOW MULTICOMPONENT CONSTRAINTS ARISE IN REACTIVE FORMULATIONS

Cornell (2002) in his detailing of MCCs (Section 4.10, p. 180), lays out this one for a metallurgy experiment that required maintaining at least 46 wt% iron in the furnace. However, five different feedstocks contained at least some iron, ranging from 20% to 60%. This limitation imposed on the total iron was maintained via the following constraint:

$$0.46 \leq 0.6x_1 + 0.5x_2 + 0.35x_3 + 0.2x_4 + 0.7x_5$$

Along similar lines, a petroleum engineer must maintain carbon within a certain range for distillation of fuel oils from many feedstocks that vary in carbon content.

The three MCCs (6, 7, and 8) presented by Snee probably stemmed from comparable requirements for maintaining a proper chemical reaction by controlling particular elements within specific bounds.

There exists everywhere a medium in things, determined by equilibrium.

—Dmitri Mendeleev (*The Principles of Chemistry*, 1891, 1, 257)

MCCs Made as Easy as Making a Pound Cake

The Pound Cake arose (pun intended) in Britain in the early 1700s ("Pound Cake History," *What's Cooking America*, https://whatscookingamerica.net/History/Cakes/PoundCake.htm). Being baked from one pound each of butter, sugar, eggs, and flour, it created a very large confection—serving multiple families.

After teaching statistical-process-control to master-bakers of Sara Lee® brand Pound Cakes (all butter, lemon, and so on), Mark resolved to experiment on the composition, not believing it is needed to be so rigidly balanced between the four basic ingredients. Perhaps a slight variation might lead to better taste.

Mark's quest for perfect Pound Cake began with research into *The Cake Bible*, first published by Rose Levy Beranbaum in 1988 (William Morrow Cookbooks). Straightaway, he learned that low-protein "cake" flour creates a desirably lighter consistency, that is, an airier texture, than all-purpose (AP) flour. However, this special flour costs about twice as much per pound as its more common AP cousin. If some, or, ideally, all, of the cake flour, could be replaced by AP flour without anyone noticing the difference, then Pound Cakes could be made far cheaper. This made a great case for application of MCCs, as you will soon see.

A DIFFERENT SPIN ON THE DANGERS OF GLUTEN IN FLOUR

Mark started off his career working for General Mills in Minneapolis, in the center of the United States' "wheat belt." During this time, he worked for some time on the startup of a starch–gluten plant, which fed off coarsely ground spring wheat from the western part of Minnesota and the vast fields of South and North Dakota. His primary objective was to increase the yield of the protein-rich gluten. If you take a handful of flour under running water, the starch runs off, leaving sticky gluten behind. Similarly, by mixing flour with water and running it over a stainless-steel bar screen, the starch goes through, while the gluten agglomerates on the surface and, in theory, rolls nicely down a chute to be dried into a high-protein powder. Unfortunately, the gluten did not always cooperate. It tended to stick together into evergrowing blobs that at times got away from Mark and his crew by rolling off the

(*Continued*)

bar-screen balcony. This created a crisis when the gluten blob got going toward the waste-water treatment plant and threatened to overwhelm it with protein. Only by the last stand of push-broom wielding workers was the disaster averted.

> I think you should send us the biggest transport plane you have, and take this thing to the Arctic or somewhere and drop it where it will never thaw.
>
> **—Lieutenant Dave, *The Blob*, 1958, Source: IMDB http://www.imdb.com/title/tt0051418/quotes**

Mark laid out these ingredients and ranges for the Pound-Cake mixture design, all in ounces (oz.) by weight:

1. Cake flour, 3–5 oz
2. AP flour, 3–5 oz
3. Sugar, 3–5 oz
4. Butter, 3–5 oz
5. Eggs, 3–5 oz

He kept the total 16 ounces, that is, one pound (1 lb), which, being one-fourth the size of the traditional Pound Cake, fit nicely into four-cavity, nonstick, mini-loaf pans for baking in his kitchen oven. But most importantly, to avoid an overdose of flour, Mark specified the following MCC:

$$3 \le A + B \le 5$$

This created a little experiment with flour types (cake versus AP) within the greater experiment on the recipe for Pound Cake (relative amounts of flour, sugar, butter, and eggs). Table 6.1 lays out 12 out of 24 mixtures from an augmented I-optimal design to fit a quadratic model. It provides 15 blends to fit the model, plus 4 lack-of-fit points (check blends) and 5 replicates (for pure error estimation). The total of 24 runs breaks down conveniently into 6 four-cavity pans that can be made in one oven batch.

Notice from the selected runs under columns A and B in Table 6.1 how the MLC keeps the total of the cake and AP flour within the bounds of 3–5 oz. Success!

Table 6.1 Selected Runs (Weights in Ounces) from Pound-Cake Experiment-Design

ID	*Build Type*	*A: Cake*	*B: AP Flour*	*C: Sugar*	*D: Butter*	*E: Eggs*
1	Model	2.5	2.5	5.0	3.0	3.0
2	Model	0.0	5.0	4.0	4.0	3.0
3	Replicate	4.0	0.0	5.0	4.0	3.0
3	Model	4.0	0.0	5.0	4.0	3.0
4	Model	5.0	0.0	3.0	5.0	3.0
5	Lack of fit	1.5	1.5	5.0	5.0	3.0
~						
14	Model	0.0	5.0	3.0	3.0	5.0
15	Model	4.0	0.0	4.0	3.0	5.0
16	Lack of fit	1.5	1.5	5.0	3.0	5.0
17	Model	4.0	0.0	3.0	4.0	5.0
18	Model	0.0	3.0	4.0	4.0	5.0
19	Lack of fit	1.5	1.5	3.0	5.0	5.0

RESULTS FROM THE POUND-CAKE EXPERIMENT

Although our job is done for laying out a case for MLCs for substituting materials, you may be curious to know what Mark came up with for an ideal recipe. The published* one calls for only AP flour (4.3 oz), with lots of sugar (4.9 oz), a generous amount of butter (4.3 oz) and just enough eggs (3.8 oz) to make up the remainder of the 16-ounce Pound Cake. This hits the spot of taste, density, and color, albeit falling well short of Sara Lee for overall scrumptiousness, heft, and appearance. A recipe that maximizes the taste alone is provided in Table 6.2. It minimizes the eggs and flour (AP only) while maximizing butter and sugar. Yum!

By the way, neither recipe requires any of the expensive cake flour. That's a bonus!

*Anderson, M., and P. Whitcomb. Mixing it up with computer-aided design. *Today's Chemist at Work*, November 1997b, p. 34.

(*Continued*)

Table 6.2 Recipe for Pound Cake That Maximizes Taste

Flour (AP Only)	3 oz
Sugar	5 oz
Butter	5 oz
Eggs	3 oz

Ratio Constraints

In some mixture problems, the ratios of components must be carefully controlled. For example, bakers of bread do well by blending a ratio of 5 parts of flour to 3 parts of water (Ruhlman, 2009). In some cases, these ratios relate to the ideal stoichiometry for chemical reactions, such as the air-to-fuel ratio in a combustion engine, which, depending on the grade of gasoline, runs at about 15-to-1 ["Air-fuel Requirement in SI Engines (Automobile)]," what-when-how, (http://what-when-how.com/automobile/air-fuel-requirement-in-si-engines-automobile/). These ratios generally can be mathematically converted to MLCs as you will see now via a real-life example.

An adhesive's chemist ran an optimal mixture design to model and control gel time for a liquid epoxy while maintaining other functional properties (Roesler, 2004). He varied three key components in a mixture, totaling 100% by weight:

1. Amine cross-linker: 20%–100% by weight
2. Plasticizer: 0%–70%
3. Catalyst: 0%–35%

The chemist imposed an additional constraint: The catalyst to plasticizer does not exceed a 4:1 ratio, that is, $C/B \leq 4$. This can be rearranged into an additive (non-ratio) MLC by multiplying both sides by A and subtracting one side from the other:

$$C \leq 4B$$

$$0 \leq 4B - C \text{ or, alternatively, } -4B + C \leq 0$$

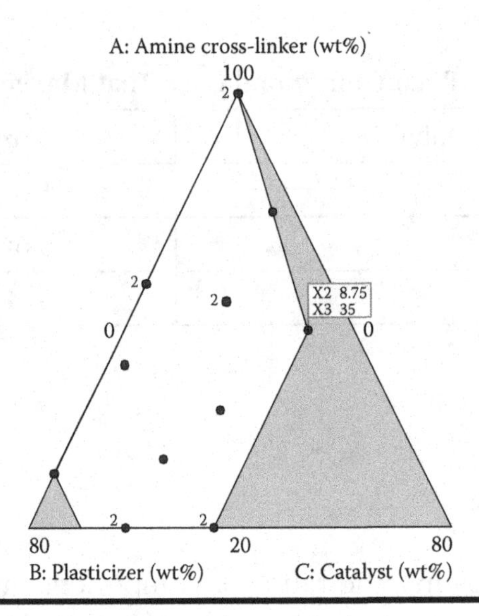

Figure 6.3 Adhesive experiment with ratio constraint.

An experiment fitting within these individual component constraints and the MLC is pictured in Figure 6.3. It was customized via I-optimal point exchange to fit a quadratic mixture model.

The edge at the upper right is flagged to show that this run (35 and 8.75 for X2 and X3, respectively) marks the edge of the feasible region where the B/C ratio hits its limit of 4.

Appendix 6A: Combining Components

Components having similar effects are not unusual, particularly when an experiment, facilitated by multicomponent constraints, includes substitutes. In that case, combining components will simplify the modeling and make the presentation of results more compelling. This turned out to be the case for the Pound-Cake experiment where two flours came into play. To illustrate why combining the flours makes sense and how this clarifies the picture, it will be helpful to introduce the "trace" plot. This plot, which we will feature in Chapter 8 on screening designs, displays the

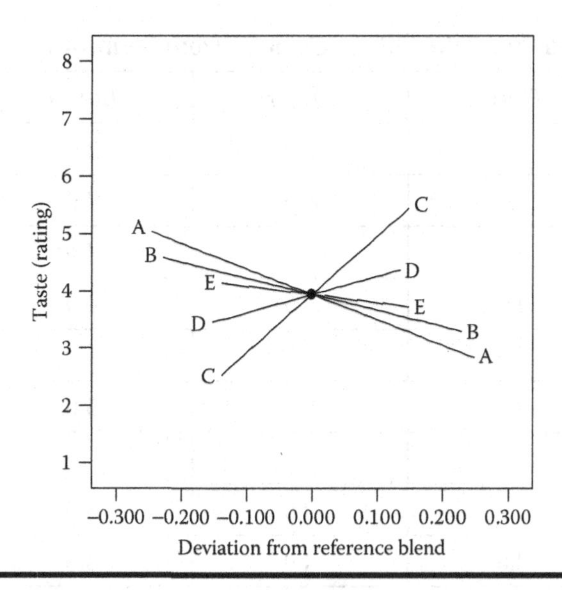

Figure 6A.1 Trace plot of taste from a Pound-Cake experiment with both flours shown.

predicted response as any given component deviates from any chosen reference point (typically the centroid) while holding all other components in constant proportion. Figure 6A.1 displays the trace plot for the Pound-Cake experiment.

Notice how the tracks for components A and B, the two flours, run alongside each other, which makes sense from a culinary perspective and it was what Mark expected to find. The same overlay of flours occurred for other attributes he measured—density and color. Seeing his hypothesis that flours make little difference confirmed, he then combined the two components into one, as shown in Table 6A.1.

This simplifies the story as illustrated by the revised trace plot in Figure 6A.2 (we expanded the axis to make the differing impacts of each main ingredient more obvious).

Now it becomes crystal clear that more sugar drives up the taste. Furthermore, flour, regardless of its type, is a "downer," as well as eggs, but to a lesser extent. Butter falls in a neutral zone, but, perhaps it may be good to be somewhat liberal in its addition to the recipe.

Table 6A.1 Selected Runs (Weights in Ounces) from Pound-Cake Experiment-Design

ID	*Flour*	*Sugar*	*Butter*	*Eggs*
1	5.0	5.0	3.0	3.0
2	5.0	4.0	4.0	3.0
3	4.0	5.0	4.0	3.0
3	4.0	5.0	4.0	3.0
4	5.0	3.0	5.0	3.0
5	3.0	5.0	5.0	3.0
~				
14	5.0	3.0	3.0	5.0
15	4.0	4.0	3.0	5.0
16	3.0	5.0	3.0	5.0
17	4.0	3.0	4.0	5.0
18	3.0	4.0	4.0	5.0
19	3.0	3.0	5.0	5.0

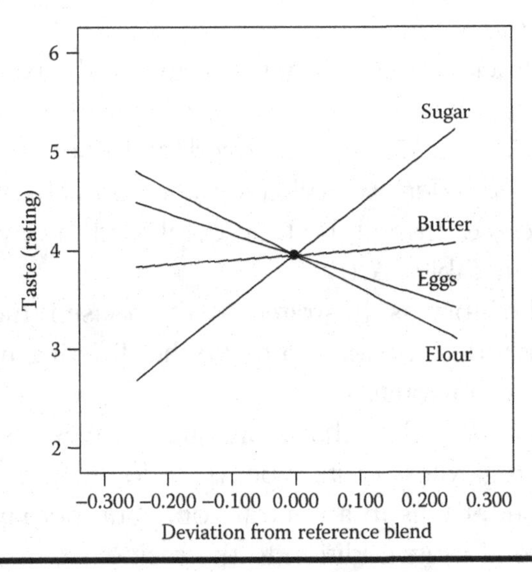

Figure 6A.2 Trace plot redone after combining the two flours.

Keep in mind that this trick of combining components only comes into play in the final analysis as a simplification. It should be done judiciously on the basis of subject-matter-knowledge, that is, not on disparate materials that coincidentally create similar effects, but only on ones that are chemically similar.

Chapter 7

Multiple Response Optimization Hits the Spot

> If your sweet spot is in the right place, there will be no vibration through the handle, and you will hear the sweetest sound in cricket, the Crack! of leather on willow.
>
> **—Blackledge (2012)**

Formulating a functional product can be extremely challenging, due to competing specifications. For example, consider a coating that must be soft to the touch, yet resistant to abrasion. Paint chemists find that reducing the amount of cross-linker achieves a velvety feel. However, it significantly degrades the durability of the coating (Spagnola et al., 2016). By deploying a mixture design that provides adequate models for critical attributes (such as the aforementioned), formulators often discover a region where components combine to meet all specifications. This is known as the "sweet spot." A simple, yet effective, way to gauge when you hit this spot for multiple responses is by scaling them all to one objective function called "desirability."

Desirability Simplified

Being on a roll with paint (pun intended), let's see how desirability works by way of a case study from the coatings industry. In this case (Anderson and Whitcomb, 1997a), experimenters explored the impact of three rheology modifiers ("RM"s) on the viscosity and flow of an architectural coating.

Each of the three components was varied from zero to one-hundred percent (0%–100%) via a second-degree simplex-lattice design augmented with the overall centroid and axial check blends. The three vertices were replicated to provide a measure of pure error. Table 7.1 lays out all of these blends and the test results. Cost is also included based on a hypothetical spread of pricing per kilogram of $7.50, $10.00, and $15.00 for RMs A, B, and C, respectively.

Fitting the rheological data to Scheffé polynomials produces these highly significant models, quadratic and linear, respectively:

Table 7.1 Experiment on Mixtures of Rheology Modifiers

ID	*Space Type*	*A:RM-A (wt%)*	*B:RM-B (wt%)*	*C:RM-C (wt%)*	*Visc Poise*	*Flow Units*	*Cost ($/kg)*
1a	Vertex	100.00	0.00	0.00	0.76	8	5.00
1b	Vertex (rep)	100.00	0.00	0.00	0.75	8	5.00
2	Center edge	50.00	50.00	0.00	1.40	7	7.50
3	Center edge	50.00	0.00	50.00	0.55	8	10.00
4a	Vertex	0.00	100.00	0.00	4.10	4	10.00
4b	Vertex (rep)	0.00	100.00	0.00	4.40	4	10.00
5	Center edge	0.00	50.00	50.00	0.90	7	12.50
6a	Vertex	0.00	0.00	100.00	0.42	9	15.00
6b	Vertex (rep)	0.00	0.00	100.00	0.40	10	15.00
7	Axial check	66.67	16.67	16.67	0.80	7	7.50
8	Axial check	16.67	66.67	16.67	1.70	7	10.00
9	Axial check	16.67	16.67	66.67	0.55	8	12.50
10	Centroid	33.33	33.33	33.33	0.80	8	10.00

$$\text{Viscosity} = 0.77\,A + 4.21\,B + 0.423\,C - 4.33\,AB + 0.31\,AC - 5.6\,BC$$

$$\text{Flow} = 7.97\,A + 4.57\,B + 9.37\,C$$

The cost is calculated deterministically via a linear equation based on the composition of each blend.

A BRIEFING ON RHEOLOGICAL PROPERTIES OF PAINT

Rheology studies the flow of liquids, primarily via measurement of viscosity. For fluids that become thinner or thicker under shear (a behavior known as non-Newtonian), other attributes must be characterized, in this case, the "process by which a wet paint flows out to a smooth film free of irregularities, that is, brush marks and ridges."* The units for flow, shown in Table 7.1 provide an indication of brush-out leveling on a scale of 1 to 10—the higher the better. For details on the actual measurement of flow, done via ASTM D4062 "Standard Test Method for Leveling of Paints by Draw-Down Method," see www.leneta.com/leveling-test.html.

> Rheology modifiers reduce dripping and spattering of paint during roller or brush application. Sag resistance of paint is improved by a rapid but controlled viscosity increase after application. During transport and storage of the paint, the rheology modifiers prevent sedimentation of the pigments within a formulation.
>
> **—Practical guide to rheology modifiers. BASF. www.dispersions-pigments.basf.com/portal/load/fid793184/BASF%20Rheology%20Modifiers%20Practical%20Guide.pdf**

**Source:* Private correspondence to authors on 5/14/01 from *Coatings Chemist Sol Vincent*, Rohm and Haas, Spring House, PA.

These models for viscosity and flow provide complete control over the rheology required for any application needs. For example, assume that the following specifications must be met:

1. Viscosity: 0.5 to 0.7 poise (target 0.6)
2. Flow: 8.0 units or better on a 10-point scale
3. Cost: $10.00/kg or less ($5.00/kg the cheapest possible)

Figure 7.1a–c shows contour plots for viscosity, flow and cost, respectively. Each of the three plots displays two or more blend-points that fall within specification. However, none of them meet all the requirements, the center of edge A–C blend—ID #5 at composition (50, 0, 50) being a near miss.

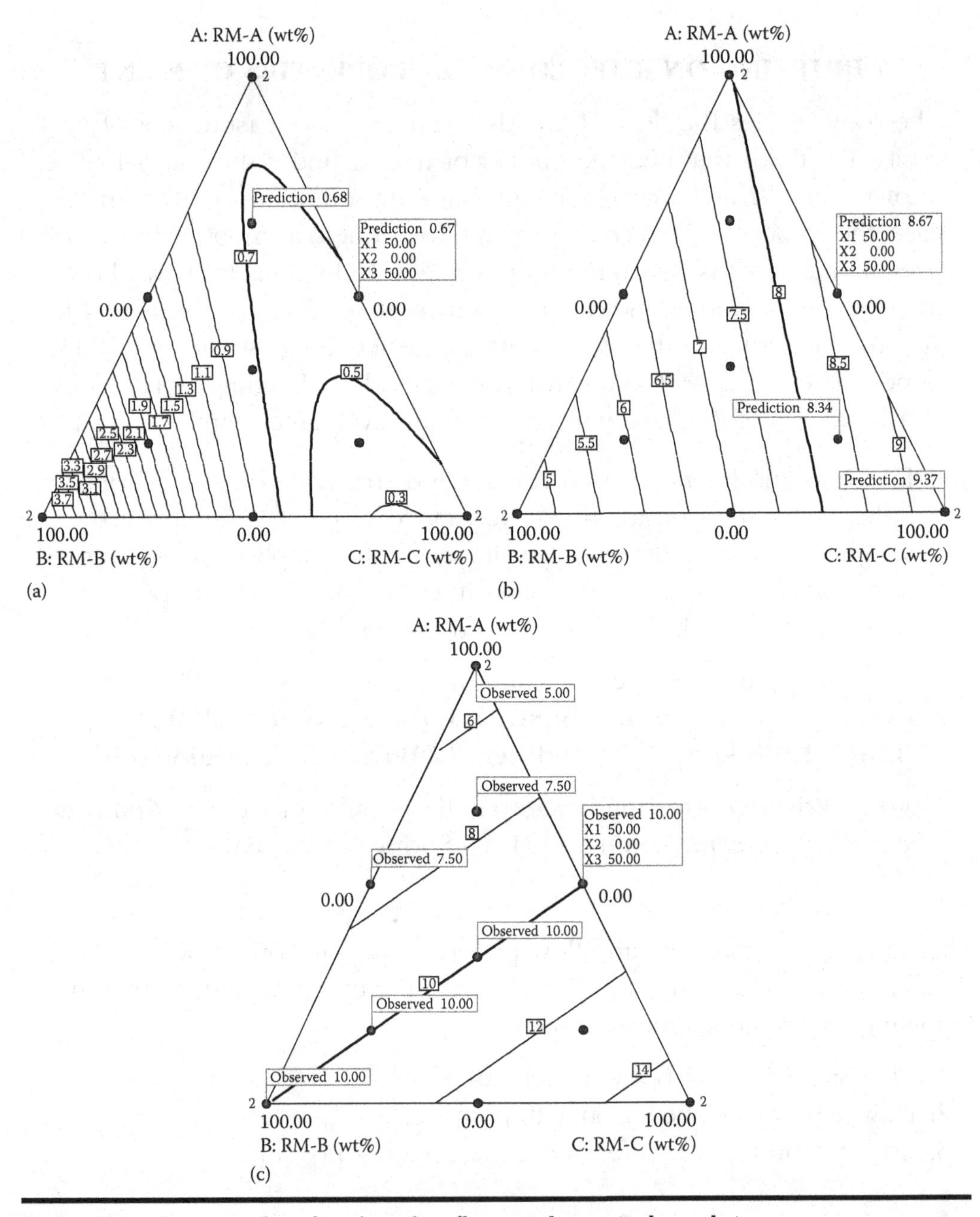

Figure 7.1 Contour plots for viscosity, flow, and cost (a, b, and c).

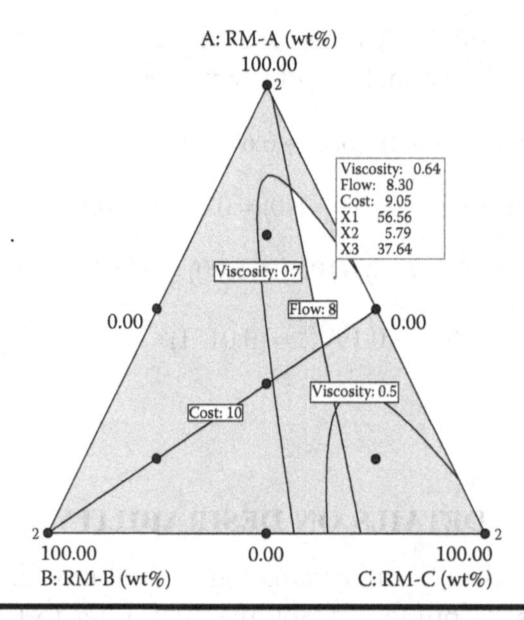

Figure 7.2 Overlay plot, showing the sweet spot for a blend of rheology modifiers.

The area meeting both specifications becomes far clearer by overlaying the three plots and then shading out the regions going out of bounds. As shown in Figure 7.2, this overlay plot provides a window that frames the sweet spot.

The flag marks a better blend than any made for the experiment: 56.56% RM-A, 5.79% RM-B, and 37.64% RM-C. (As a practical matter, the coatings chemist would do well by rounding these levels to 56.5%, 5.8%, and 37.7%, respectively.) This optimal blend beats mixture #5 on two out of the three specifications—viscosity (a bit closer to target) and cost (much cheaper). Only on flow does it fall off somewhat—8.30 versus 8.67 for blend #5.

To achieve this optimal blend, all three response measures were rescaled to one objective function called "desirability" going from 0 (none) to 1 (complete). Figure 7.3 shows how desirability ramps up and down for viscosity (a target), up for flow (maximized) and down for cost (minimized). (The number lines are bounded by the observed extremes printed in smaller font size.)

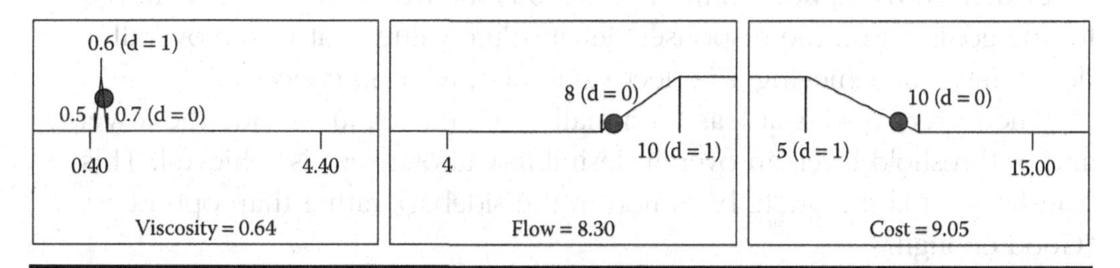

Figure 7.3 Ramps view of the most desirable blend of rheology modifiers.

An individual's desirability (small case "d") and the overall desirability (upper case "D") at this optimal blend point are:

$$d_1 = (0.70 - 0.64)/(0.70 - 0.60) = 0.6$$

$$d_2 = (8.3 - 8.0)/(10.0 - 8.0) = 0.3/2.0 = 0.15$$

$$d_3 = (10.00 - 9.05)/(10.00 - 5.00) = 0.95/5.00 = 0.19$$

$$D = (0.6 \times 0.15 \times 0.19)^{1/3} = (0.0171)^{1/3} = 0.26$$

DETAILS ON DESIRABILITY

For mathematical details on calculating desirability and the attendant search algorithms for numerical optimization, refer to Chapter 6 of *RSM Simplified, 2nd ed.* There, you will also find a background on how to prioritize responses by scaling them in terms of "importance." For example, in this case, a coatings chemist could push the selection of the most desirable blend of rheology modifiers closer to the target on viscosity by making the response most important—5 on the 5-point scale, while setting both flow and cost at the least important level of 1.

> Whereas economic man maximizes, selects the best alternative from among all those available to him, his cousin, administrative man, satisfices, look for a course of action that is satisfactory or "good enough.
>
> **—Simon (1997)**

As evidenced by Figure 7.4, the D of 0.26 is the best that can be achieved for the goals set on the responses. Its absolute value, that is, the overall desirability not achieving a perfect value of 1, is of no concern.

When every goal is at least minimally met, that is, just inside the lower and or threshold level, an overall desirability above zero is achieved. This "satisfices" (a la the quote by Simon in the sidebar), rather than optimizes. "Good enough."

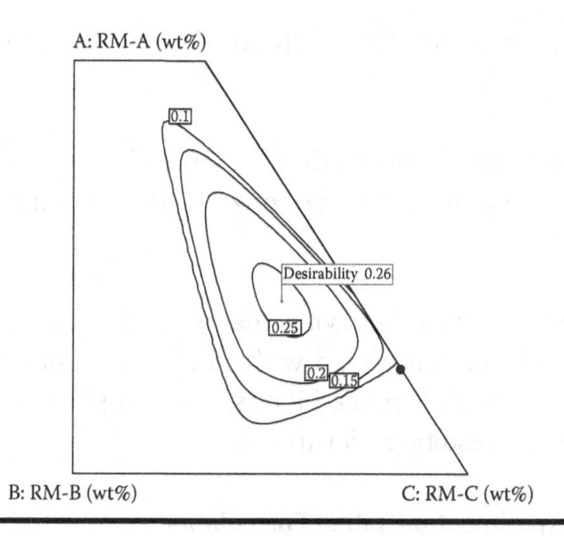

Figure 7.4 Contour plot of desirability (magnified view).

Framing the Sweet Spot and Narrowing It Down to Achieve Quality by Design (QbD)

The overlay plot provides a window into regions where experimenters can meet specifications for multiple responses. Thus, it facilitates the development of a QbD "design space," which is defined by the Food and Drug Administration (FDA) as the "multidimensional combination and interaction of input variables (e.g., material attributes) and process parameters that have demonstrated to provide assurance of quality" (Guidance for Industry, 2009). However, operating at the outer edges of the overlay window, that is, its frame, cannot assure quality due to it being only a 50/50 chance of that response going out of bounds. To improve these odds, we advise you impose a confidence interval (CI) or, better yet, a tolerance interval (TI). The following case study illustrates the application of the CI and, ultimately the TI, to achieve a QbD design space.

Pharmaceutical formulators sought the sweet spot for the following components in a hydrophilic tablet with the ranges listed by weight percent:

1. Lactose, 5%–42%
2. Phosphate, 5%–47%
3. Cellulose, 5%–52%
4. Polymer, 17%–25%
5. Drug, 1%–2%

They wanted their final formula to achieve the following two specifications reliably:

1. Dissolution time (T) in hrs to dissolve 50%, 5–11 (target 8)
2. Hardness in kiloponds (kP), maximize from 4, at least, with 8 being good enough

Assuming quadratic mixture behavior, that is, nonlinear blending effects, the chemists set up an I-optimal design with 15 model points. They augmented this base experiment with 5 check-blends, and 5 replicates chosen optimally. Table 7.2 lays out the resulting 25 runs.

Table 7.2 QbD Experiment on Tablet Formulation

ID	*A: Lact (wt%)*	*B: Phos (wt%)*	*C: Cell (wt%)*	*D: Poly (wt%)*	*E: Drug (wt%)*	*Dis @ t (50%) hrs*	*Hard kP*
0	23.33	25.83	28.33	21.00	1.50	8.23	7.56
1	42.00	20.00	20.00	17.00	1.00	6.27	8.24
2	22.33	5.00	52.00	19.67	1.00	6.14	6.48
3	5.00	47.00	26.00	21.00	1.00	11.61	9.44
4a	42.00	29.67	5.00	22.33	1.00	7.89	8.30
4b	42.00	29.67	5.00	22.33	1.00	8.42	6.43
5	5.00	32.00	37.00	25.00	1.00	14.12	3.97
6	29.67	33.42	16.67	19.00	1.25	8.18	7.35
7	20.17	15.42	40.17	23.00	1.25	8.99	4.13
8	29.67	47.00	5.00	17.00	1.33	6.33	6.09
9	42.00	5.00	29.33	22.33	1.33	5.69	8.69
10a	5.00	24.33	52.00	17.00	1.67	7.56	6.61
10b	5.00	24.33	52.00	17.00	1.67	8.27	6.48
11	14.17	36.42	24.67	23.00	1.75	9.16	8.39
12a	17.00	47.00	17.00	17.00	2.00	6.02	7.22
12b	17.00	47.00	17.00	17.00	2.00	4.66	6.10
13a	30.00	5.00	46.00	17.00	2.00	4.53	8.04

(*Continued*)

Table 7.2 (*Continued*) QbD Experiment on Tablet Formulation

ID	*A: Lact (wt%)*	*B: Phos (wt%)*	*C: Cell (wt%)*	*D: Poly (wt%)*	*E: Drug (wt%)*	*Dis @ t (50%) hrs*	*Hard kP*
13b	30.00	5.00	46.00	17.00	2.00	5.61	9.17
14a	42.00	30.00	5.00	21.00	2.00	3.94	6.20
14b	42.00	30.00	5.00	21.00	2.00	4.56	7.79
15	22.67	40.00	10.33	25.00	2.00	7.98	8.86
16	5.00	47.00	21.00	25.00	2.00	10.45	10.54
17	33.33	12.00	27.67	25.00	2.00	9.21	9.47
18	29.00	5.00	39.00	25.00	2.00	7.64	9.83
19	8.67	12.33	52.00	25.00	2.00	9.84	5.09

After fitting the full quadratic model (A, B, C, D, E, AB, AC, AD, AE, BC, BD, BE, CD, CE, DE) to each response, the following terms were removed via backward elimination at p of 0.1 (for background on this tool, see "A Brief Word on Algorithmic Model Reduction", *RSM Simplified, 2nd ed.*, pp. 30–32.):

1. Dissolution: AB, AC, AE, CE, DE
2. Hardness: AC, BC, CD, DE

Based on these models, a numerical search found the optimum depicted by the desirability ramps in Figure 7.5. Factors A through D ranged freely within their initial design constraint. However, factor D—the drug—was set at 1% for this optimization, that is, the low dosage of active pharmaceutical ingredient (API). This is an equality constraint. It can be applied only to factors when searching for most desirable formulations (Figure 7.5). (A separate optimization was done for the 2% API tablet.)

This most desirable formulation for the 1% API tablet is flagged on the overlay plot of Figure 7.6a (first plot or plot on the left), which depicts A, B and C with component D (polymer) sliced at its optimal level (17%).

Figure 7.6b (second plot or middle plot) shows a more conservative sweet spot framed by the CI's. The flagged formulation still falls well within boundaries, despite them being narrowed to account for the uncertainty that naturally stems from limited experimental data. However, FDA regulators want assurance that, all or nearly all, individual units (not just on average!)

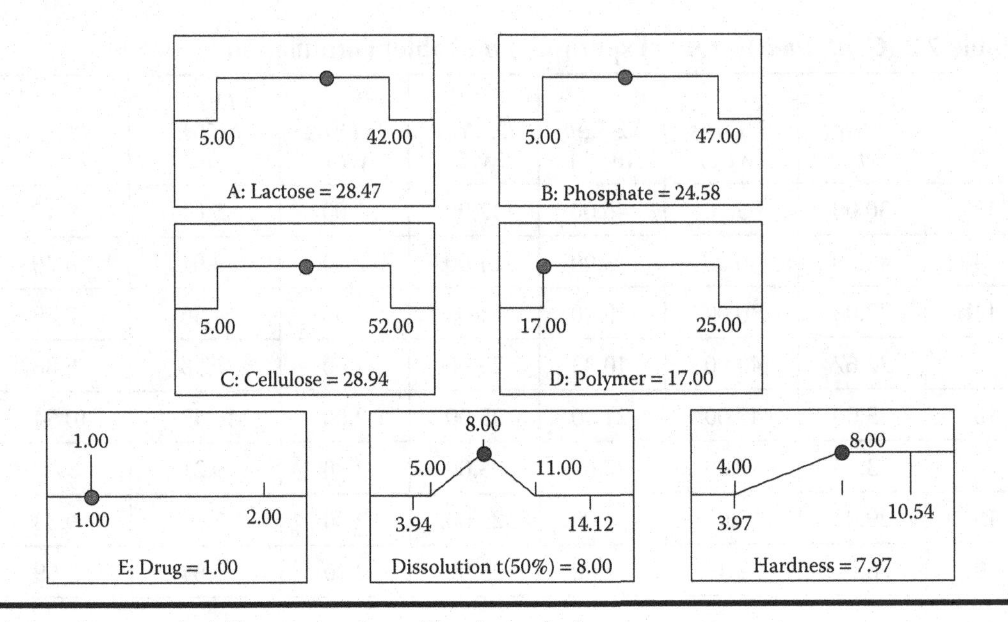

Figure 7.5 Desirability ramps for tablet formulation.

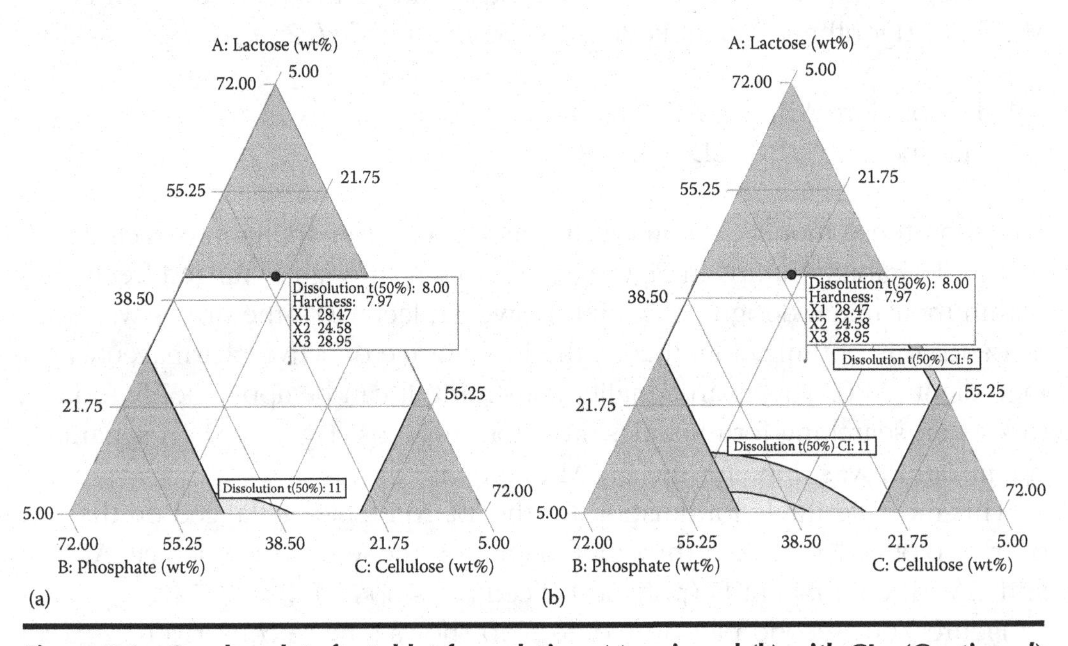

Figure 7.6 Overlay plots for tablet formulation: (a) as-is and (b) with CI. ***(Continued)***

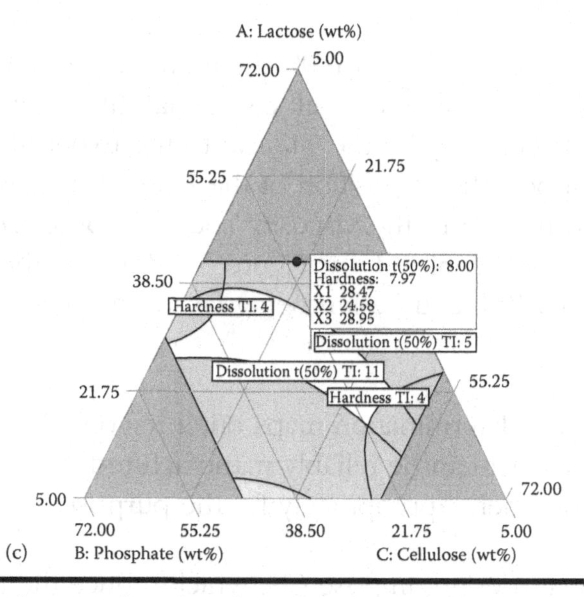

Figure 7.6 (Continued) Overlay plots for tablet formulation: (c) with TI.

will meet specification. This requires imposing a TI that provides a high level of confidence, say 95%, out of 99% (a generally acceptable threshold) of the population meet (or exceed) specifications.

Figure 7.6c (third plot or plot on the right) shows how the TI pushes the dissolution boundary well beyond that created by the CI. Also, the hardness specification now imposes a restriction. However, the flagged formulation remains acceptable. Thus, this qualifies as a QbD design space, within which the formulation can float with little risk of any tablets going out of specification.

DETAILS ON COMPUTING CI AND TI AND THEIR IMPACT ON DESIGN SIZING

To develop assurance for the quality of their products, manufacturers must back off from their specification boundary by the CI, or, more conservatively for QbD, the tolerance interval (TI). This requires computation of the intervals' half-width. For references to the required calculations, included specialized ones for designed experiments, see "Using DOE with Tolerance Intervals to Verify Specifications" (Whitcomb and Anderson, 2011). In general, the TIs widen directly with confidence level and proportion of in-specification product. Thus, the more demanding one gets for

(Continued)

confidence and/or the nearer to 100% the requirements reach, the smaller the QbD design space becomes—all else equal. In fact, we often see the space covered up entirely by the TI, due to the experiment being sized too small to support the imposition of the interval. For example, adding the TI (or even the CI) to the viscosity specification closes the window in the overlay plot depicted by Figure 7.2 for the rheology-modifier formulation—this 10-blend design being far too meager for QbD.

A QbD design-space determination maps out a region where the in-specification product can be reliably manufactured. We advise that you size your experiment appropriately for the purpose:

- For "functional" design, impose CIs, which reduce the risk of mean results falling outside the allowable operating boundaries.
- To verify that your product falls within final manufacturing specifications, for example, to achieve a QbD design space, enforce tolerance intervals (TI).

Keep in mind that TIs range far wider than CIs. Therefore, a relatively large experiment will be required to create any design space meeting such a rigorous statistical requirement.

Practice Problem

To practice using the statistical techniques you learned in Chapter 3, work through the following problems.

Problem 7.1

Follow up on the detergent case from Problem 3.1 by completing the multiple-response optimization. Is there a sweet spot for the formulation that achieves all specifications? Find out by using the computer tool specialized for this purpose—see *About the Software* for the path to the program and the link to accompanying tutorials. Download and print the *Mixture Design Tutorial* (*Part 2/2—Optimization*). Follow it through to the conclusion of this case study.

PROPAGATION OF ERROR (POE)

As detailed in the tutorial, if you enter the variation in your component amounts (e.g., due to how they are weighed out), the software can generate POE plots showing how that error transmits to the response. Look for compositions that minimize POE, thus creating a formula that's robust to variations in the measured amounts. For details on POE refer to Chapter 9 in *RSM Simplified, 2nd ed.*

Chapter 8

Screening for Vital Components

Nearly all the grandest discoveries of science have been but the rewards of accurate measurement and patient long-continued labor in the minute sifting of numerical results.

—Lord Kelvin (Presidential Address to Royal Society, 1871, quoted p. 940 in *Life of Lord Kelvin*, Silvanus Phillips Thompson, 1910)

For process troubleshooting, a tried-and-true strategy of experimentation begins with screening previously unknown factors, from which engineers often find a vital few to carry forward to the next stages—the characterization of interactions and, finally, optimization. Chemists usually jump directly into optimization when experimenting on formulations, relying on their subject-matter knowledge to narrow down the field of components. However, as this chapter will demonstrate, screening for main-component effects can uncover vital ingredients that might otherwise be overlooked.

Screening designs only fit the linear model. We recommend them for 6 or more components, for which higher order models become too big to accommodate in an affordable experiment. This becomes clear

Table 8.1 Number of Terms in Scheffé Polynomials for 5–10 Components

Components (q)	*Linear*	*Quadratic*	*Special Cubic*
5	5	15	25
6	6	21	41
7	7	28	63
8	8	36	92
9	9	45	129
10	10	55	175

from Table 8.1, which, starting with 5 components (where a similar accounting in Table 2A.1 ends), enumerates terms for mixture models of degrees 1 through 3.

Keeping in mind that when building a stout experiment design, it pays to add another 10 blends (5 to check lack of fit, plus 5 replicates), a robust 6-component design for the special cubic model comes to 51 blends (41 + 10). This goes well beyond the scope of most formulators in our experience. On the other hand, a screening experiment, aimed at estimating only the linear terms, requires at a minimum, 6 blends, and, at most, 23 blends per generally acceptable practice, which we will now detail.

We recommend that, if you wish to experiment on 6 or more components, you first screen them down and then follow up with an optimization design that models quadratic (second degree) or special cubic (third degree) behavior.

High-Octane Simplex Screening Designs

We begin the detailing of screening designs with ones that cover a simplex region, for example, when the amounts of each component vary from zero to one-hundred percent ($0 \leq x_i \leq 1$ in coded scale). This most-basic scenario will be illustrated by a case study on octane-improvement additives. The data—modified somewhat for explanatory purposes—come from Snee and Marquardt (Snee and Marquardt, 1976).

DON'T KNOCK IT

The octane number characterizes a gasoline's antiknock quality, that is, its capacity to withstand damaging premature detonation in an engine's combustion chamber. Isooctane was the original antiknock additive—hence the name "octane" for the rating. The higher the octane number, the more compression the fuel can handle. When gasoline is first distilled from oil, it has an octane number of about 70, which would severely limit engine efficiency. An octane rating of 87 meets the needs for regular gasoline, a level that correlates directly to a mixture of 87% isooctane and 13% heptane. (A 100% heptane fuel is the zero point of the octane rating scale.)

In 1921, a team of General Motors chemists discovered that tetraethyl lead worked wonders for octane ratings. However, due to health risks, the Clean Air Act of 1996 banned the sale of leaded gasoline in the U.S.A. One of the cheapest and most common replacements for the lead as an octane booster is ethanol. However, its energy content is only 70% of gasoline. Also, ethanol creates a host of engine issues. Currently, the best option to ethanol is a toxic mixture of benzene, toluene, ethyl–benzene, and xylene (BTEX).

The search for safe and effective octane-additives continues.

> A difference of only 13 points in octane number made possible the defeat of the Luftwaffe by the R. A. F. in the fall of 1940.
>
> **—Kalichevsky (1943)**

Petroleum chemists want to screen 10 octane additives that will be blended in a fixed amount to a specific quantity of gasoline. Blends of these additives will not be ruled out, but that can be studied later in a follow-up experiment. For now, the objective is to identify which, if any, components generate promising linear effects. A design that boldly spans the ranges will be most effective for this screening experiment. Figure 8.1 shows a good choice—a "simplex screening" design (Snee and Marquardt, 1976).

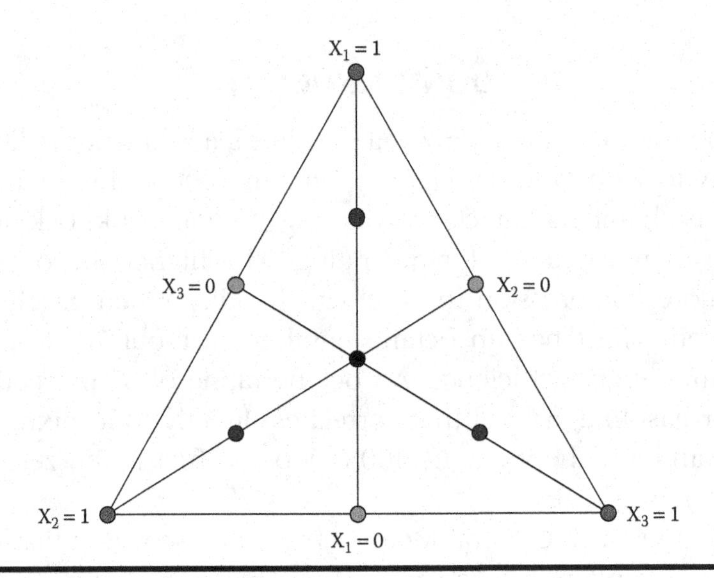

Figure 8.1 Simplex screening design.

Keep in mind that along the axes of a simplex (e.g., X_1 from 0 to 1), the proportion of each component spans its entire range while the relative proportions of the other components to one another remain unchanged. At a minimum, a simplex screening design should be comprised of the q vertices (10 in the case of the octane additives). Next, if the experimental budget allows, we advise you add the centroid and replicate it 5 or so times to provide an estimate of pure error. Beyond that, consider adding q axial points (located midway between the centroid and the vertices) to fill the gaps and provide more power. This brings the total blends to 2q + 1. Finally, another q "end effect" blend (Snee and Marquardt, 1976, p. 22) can be added—these being the points along the sides in Figure 8.1 where one component goes to zero (e.g., $X_1 = 0$). These latter blends, expanding the design to 3q + 1, measure the effect from the complete elimination of any single ingredient.

The petroleum chemists chose the middle-sized 2q + 1 simplex screening design shown in Table 8.2, which included 5 replicates of the centroid

Table 8.2 Octane-Additive Design and Results

#	$A:X_1$	$B:X_2$	$C:X_3$	$D:X_4$	$E:X_5$	$F:X_6$	$G:X_7$	$H:X_8$	$J:X_9$	$K:X_{10}$	Octane
1	1.00	0.00	0.00	0.00	0.00	0.00	0.00	0.00	0.00	0.00	62.30
2	0.00	1.00	0.00	0.00	0.00	0.00	0.00	0.00	0.00	0.00	82.90
3	0.00	0.00	1.00	0.00	0.00	0.00	0.00	0.00	0.00	0.00	86.00
4	0.00	0.00	0.00	1.00	0.00	0.00	0.00	0.00	0.00	0.00	77.50
5	0.00	0.00	0.00	0.00	1.00	0.00	0.00	0.00	0.00	0.00	79.10
6	0.00	0.00	0.00	0.00	0.00	1.00	0.00	0.00	0.00	0.00	74.90
7	0.00	0.00	0.00	0.00	0.00	0.00	1.00	0.00	0.00	0.00	80.20
8	0.00	0.00	0.00	0.00	0.00	0.00	0.00	1.00	0.00	0.00	78.30
9	0.00	0.00	0.00	0.00	0.00	0.00	0.00	0.00	1.00	0.00	77.30
10	0.00	0.00	0.00	0.00	0.00	0.00	0.00	0.00	0.00	1.00	84.40
11	0.55	0.05	0.05	0.05	0.05	0.05	0.05	0.05	0.05	0.05	74.20
12	0.05	0.55	0.05	0.05	0.05	0.05	0.05	0.05	0.05	0.05	83.60
13	0.05	0.05	0.55	0.05	0.05	0.05	0.05	0.05	0.05	0.05	83.60
14	0.05	0.05	0.05	0.55	0.05	0.05	0.05	0.05	0.05	0.05	79.00
15	0.05	0.05	0.05	0.05	0.55	0.05	0.05	0.05	0.05	0.05	79.90
16	0.05	0.05	0.05	0.05	0.05	0.55	0.05	0.05	0.05	0.05	77.90
17	0.05	0.05	0.05	0.05	0.05	0.05	0.55	0.05	0.05	0.05	78.00
18	0.05	0.05	0.05	0.05	0.05	0.05	0.05	0.55	0.05	0.05	82.50
19	0.05	0.05	0.05	0.05	0.05	0.05	0.05	0.05	0.55	0.05	79.90
20	0.05	0.05	0.05	0.05	0.05	0.05	0.05	0.05	0.05	0.55	81.90
21	0.10	0.10	0.10	0.10	0.10	0.10	0.10	0.10	0.10	0.10	75.50
22	0.10	0.10	0.10	0.10	0.10	0.10	0.10	0.10	0.10	0.10	79.50
23	0.10	0.10	0.10	0.10	0.10	0.10	0.10	0.10	0.10	0.10	78.70
24	0.10	0.10	0.10	0.10	0.10	0.10	0.10	0.10	0.10	0.10	80.30
25	0.10	0.10	0.10	0.10	0.10	0.10	0.10	0.10	0.10	0.10	81.70

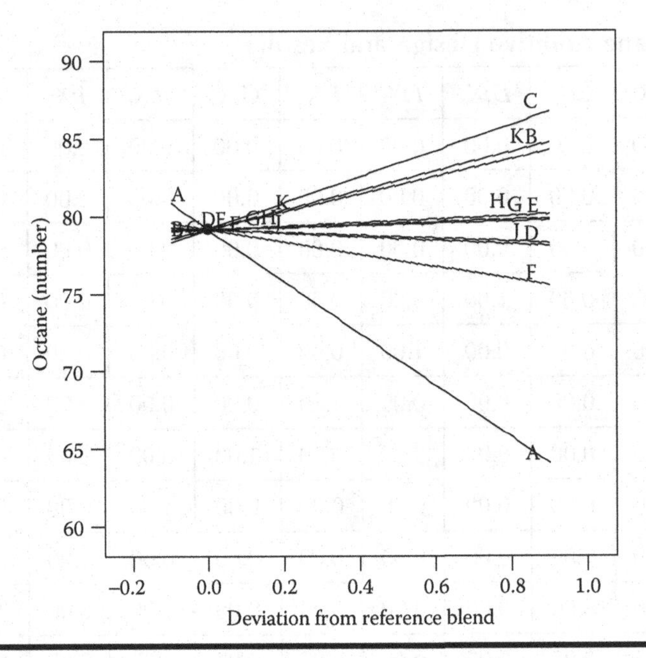

Figure 8.2 Trace plot of component effects on octane number.

(blends 21–25) and the 10 axial check blends (#s 11–20), as well as the 10 vertices (#s 1–10) that must be tested at a minimum.

The resulting octane numbers, which range from 62.3 to 86.0, produce a significant fit to the linear mixture model. The most important view of the relative effects is provided by the trace plot depicted in Figure 8.2.

As we explained in Appendix 6A, this plot displays the predicted response as any given component deviates from any chosen reference point (typically the centroid), while holding all other components in constant proportion. For example, follow component A (X_1) from the nexus of all 10 ingredients at the zero point on the abscissa (0.0 deviation from reference blend). As indicated by the line going up to the left, reducing A to its minimum of zero (−0.1 deviation), that is, taking it out of the mixture, increases the octane. As more and more of component A goes into the blend, it steadily degrades the octane. The lowest response prediction comes when the additive is comprised only of A (0.9 units above the centroid reference blend). Clearly, ingredient X_1 (component A) should be discarded.

VARIATIONS IN HOW TO DRAW A TRACE PLOT

As detailed in the Appendix, two methods prevail for the direction of the trace—Cox versus Piepel. In this case, because all components go from zero to one-hundred percent (0%–100%), the plots are identical. Otherwise, though, you will see differences between the Cox and Piepel traces that can lead to ambiguous interpretation for all but very active components.

PS: Perhaps you noticed that letter "I" is missing from the design layout in Table 8.2 and the trace plot—the components skip from "H" to "J." This is done because "I" is reserved as a symbol for the intercept in conventional polynomial models.

On a positive note, components C (X_3), K (X_{10}) and B (X_2) increase octane numbers dramatically—they are keepers. The other 6 components create little effect, one way or the other, so depending on the whims of the petroleum chemists, they can be kept or tossed. These subject-matter experts may also make something out of those components whose responses follow each other very closely, for example, J (X_9) and D (X_4)—perhaps they are similar chemically or the same, but from differing suppliers. That might be good to know. Furthermore, such ingredients could be combined to simplify follow up experiments.

Measuring the Effect of a Component

Although many inferences can be made from a trace plot, it will not suffice for those who like numerical results. To satisfy these "quants," effects (E) can be measured by the following equation, which takes the difference from the model coefficient of any individual (i) component to the average of all the others:

$$E_i = \beta_i - \frac{\sum_{j \neq i}^{q} \beta_j}{(q-1)}$$

Table 8.3 Octane-Additive Coefficients and Effects

Comp	*Beta*	*Effect*	*p-Value*
A-X_1	64.03	–16.81	<0.0001
B-X_2	84.27	5.68	0.0071
C-X_3	86.75	8.43	0.0003
D-X_4	78.11	–1.17	0.5307
E-X_5	79.75	0.65	0.7248
F-X_6	75.59	–3.97	0.0457
G-X_7	79.87	0.79	0.6719
H-X_8	80.15	1.10	0.5557
J-X_9	78.31	–0.95	0.6108
K-X_{10}	84.79	6.25	0.0037

where ß (beta) represents the model coefficients, q is the number of components and j is an index for all the other ingredients aside from the one being measured for effect. Table 8.3 provides the linear model coefficients for the 10 octane additives as well as their linear effect. For good measure, p-values are also displayed.

In this case, the coefficients are simply the predicted octane number for the gasoline with only that additive. For example, component A (X_1) can be expected to produce a meager octane of 64.03. Its effect computes extremely negative:

$$E_A = 64.03 - 80.84 = -16.81$$

where 80.84 is the octane number on average for the nine other additives.

Extreme Vertices Design for Non-Simplex Screening

When component constraints violate the simplex geometry, we recommend extreme-vertices screening design (EVD), which comprises as follows:

- 2q vertices chosen optimally
- n replicates (5 generally suffices) of the centroid.

Because some components in a nonsimplex design change less than others, their effects must be adjusted accordingly and it becomes helpful to look at their gradient. We will detail this by the following case study drawn in most part (some changes made for the sake of simplicity) from Snee and Marquardt (1976, p. 26).

Chemists crafting a formulation came up with eight candidate components that can provide a key attribute—the higher, the better—for a new product. They are determined to screen them down to a vital few. Due to varying potencies, the ranges must differ as follows:

1. X_1, 0.10–0.45
2. X_2, 0.05–0.50
3. X_3, 0.00–0.10
4. X_4, 0.00–0.10
5. X_5, 0.10–0.60
6. X_6, 0.05–0.20
7. X_7, 0.00–0.05
8. X_8, 0.00–0.05

These ranges do not form a simplex. Therefore, an EVD is required. The effects of the components are unknown. With the aid of a computerized algorithm, 182 vertices are identified as extremes in the seven-dimensional (q − 1) mixture space. The program applies the D-optimal criterion to select the best 16 vertices (2q). It then locates the centroid and replicates it 5 times at random intervals in the resulting 21-blend layout for the experiment. The resulting EVD is laid out in Table 8.4 with the vertices first and then the 5 replicates of the centroid (rounded to nearest one-hundredth for the sake of convenience).

Figure 8.3 displays the trace plot for the eight-component ESD. It's magnified to give a better view of the four highly-restricted components—C (X_3), D (X_4), G (X_7), and H (X_8). Notice that the other, relatively wide-ranging components go off the chart at left and/or right.

Table 8.4 Eight-Component EVD

Std	*A:X_1*	*B:X_2*	*C:X_3*	*D:X_4*	*E:X_5*	*F:X_6*	*G:X_7*	*H:X_8*	*Y*
1	0.10	0.05	0.10	0.10	0.55	0.05	0.00	0.05	118
2	0.15	0.50	0.00	0.00	0.10	0.20	0.00	0.05	17
3	0.10	0.50	0.10	0.00	0.20	0.05	0.00	0.05	34
4	0.10	0.50	0.00	0.00	0.10	0.20	0.05	0.05	29
5	0.10	0.05	0.00	0.00	0.55	0.20	0.05	0.05	114
6	0.45	0.15	0.10	0.10	0.10	0.05	0.05	0.00	11
7	0.10	0.50	0.00	0.00	0.20	0.20	0.00	0.00	22
8	0.45	0.05	0.10	0.05	0.10	0.20	0.00	0.05	21
9	0.45	0.05	0.00	0.00	0.40	0.05	0.00	0.05	42
10	0.20	0.50	0.00	0.10	0.10	0.05	0.00	0.05	13
11	0.10	0.05	0.00	0.10	0.55	0.20	0.00	0.00	79
12	0.10	0.10	0.10	0.00	0.60	0.05	0.05	0.00	115
13	0.45	0.05	0.00	0.00	0.45	0.05	0.00	0.00	48
14	0.45	0.20	0.00	0.10	0.10	0.05	0.05	0.05	20
15	0.45	0.10	0.10	0.00	0.10	0.20	0.05	0.00	9
16	0.10	0.50	0.00	0.10	0.20	0.05	0.05	0.00	29
17	0.24	0.22	0.05	0.05	0.28	0.12	0.02	0.02	35
18	0.24	0.22	0.05	0.05	0.28	0.12	0.02	0.02	54
19	0.24	0.22	0.05	0.05	0.28	0.12	0.02	0.02	35
20	0.24	0.22	0.05	0.05	0.28	0.12	0.02	0.02	42
21	0.24	0.22	0.05	0.05	0.28	0.12	0.02	0.02	37

Components E (X_5), G (X_7), and H (X_8) stand out as "keepers" by their steeply upward slopes. Table 8.5 lists the gradients, calculated by dividing the component effect by the difference between the highest and lowest component value on the trace. They reinforce the value of E, G and H, while, on the other hand, accentuating the deleterious impacts of components A (X_1) and B (X_2).

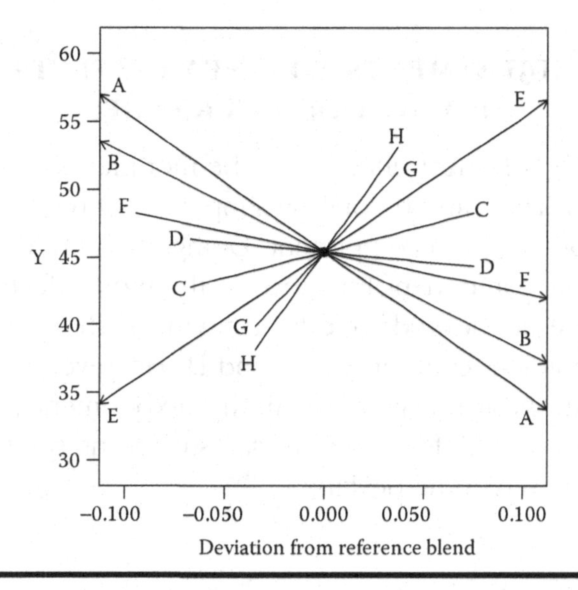

Figure 8.3 Trace plot of component effects from eight-component EVD case.

Table 8.5 Gradients for Eight-Component EVD

Comp	*Gradient*	*p-Value*
A-X_1	–101.21	<0.0001
B-X_2	–70.02	<0.0001
C-X_3	40.27	0.1664
D-X_4	–12.39	0.6655
E-X_5	102.05	<0.0001
F-X_6	–28.66	0.1321
G-X_7	160.75	0.0096
H-X_8	210.55	0.0015

Components C (X_3), D (X_4), and F (X_6) do not affect the response much one way or the other, as evidenced by their insignificant ($p > 0.05$) gradients. If any or all of these ingredients come cheap, these might be retained in the formula. Otherwise, they should be taken out.

ADJUSTMENTS TO LINEAR EFFECTS DUE TO VARYING RANGES

The interpretation of linear effects must be modified to account for complex constraints. This can be done by simply adjusting each effect by the relative range of the given component. Doing so in the eight-component EVD case reveals component E as, by far, the most effective for increasing the response (as desired) due to it having such a wide range, relative to higher-gradient components G and H. However, the gradients tell the story best in these nonsimplex mixture experiments, and they keep things simple, so we will focus on this statistic for interpretation of results from nonsimplex screening designs.

Moving on from here, the formulators will do well by a follow-up optimization experiment on components E, G, and H, aimed at detecting nonlinear blending effects, for example, via an I-optimal custom design for a quadratic or special-cubic model. If the ranges on G and H can be expanded without dire consequence, that will be all the better.

A word of caution—the cases we made for screening only included one response, whereas in real life a formulation generally must provide many functions. When you have multiple responses like this, things get far more complicated when trying to eliminate ingredients, because any one may be good for one attribute, while bad for another. It may be difficult to find a single component that is entirely superfluous. Nevertheless, when recipes become bloated or many ingredients can be considered, a screening design might be just the ticket.

Appendix 8A: Trace Plots—Cox versus Piepel Direction

The trace plot shows the effects of changing each component along an imaginary line from the reference blend (defaulted to the overall centroid) to the vertex. In the Cox direction (Cox, 1971), done in the real-coded space, as the amount of this component increases, the amounts of all other components decrease, but their ratio to one another remains constant.

Chemists may like this because it preserves the reaction stoichiometry. However, when plotted in this direction, traces of highly constrained mixture components, such as a catalyst for a chemical reaction, become truncated. Thus although it no longer holds actual ratios constant, we feel that Piepel's direction (Piepel, 1982), done in pseudo-coded space, provides a more helpful plot by providing the broadest coverage of the experimental space. For example, Figure 8.3 is done by the Piepel method, as well as the associated gradients in Table 8.5.

How these two options for trace plots differ becomes clear with a few pictures from an experiment to adulterate an acrylonitrile–butadiene–styrene (ABS) plastic pipe with otherwise unwanted coal-tar pitch (Koons and Wilt, 1982). The other two ingredients (besides the pitch) were grafted polybutadiene (graft) and styrene–acrylonitrile (SAN). Table 8A.1 lays out the experiment—a second-degree simplex lattice augmented with the centroid and axial check blends (rounded a bit for the convenience of the formulation).

Figure 8A.1a and b shows how the directions of Cox and Piepel diverge. Both begin at the same point—the centroid. Cox goes from there to the

Table 8A.1 Pipe Experiment

Std	*A: Graft* (*Wt%*)	*B: SAN* (*Wt%*)	*C: Pitch* (*Wt%*)	*Izod* (*ft-lb/in*)
1	70.0	30.0	0.0	7.4
2	45.0	55.0	0.0	6.1
3	45.0	30.0	25.0	0.8
4	57.5	42.5	0.0	7.3
5	57.5	30.0	12.5	3.9
6	45.0	42.5	12.5	2.1
7	60.0	35.0	5.0	6.1
8	50.0	45.0	5.0	4.9
9	50.0	35.0	15.0	2.5
10	53.5	38.0	8.5	4.4

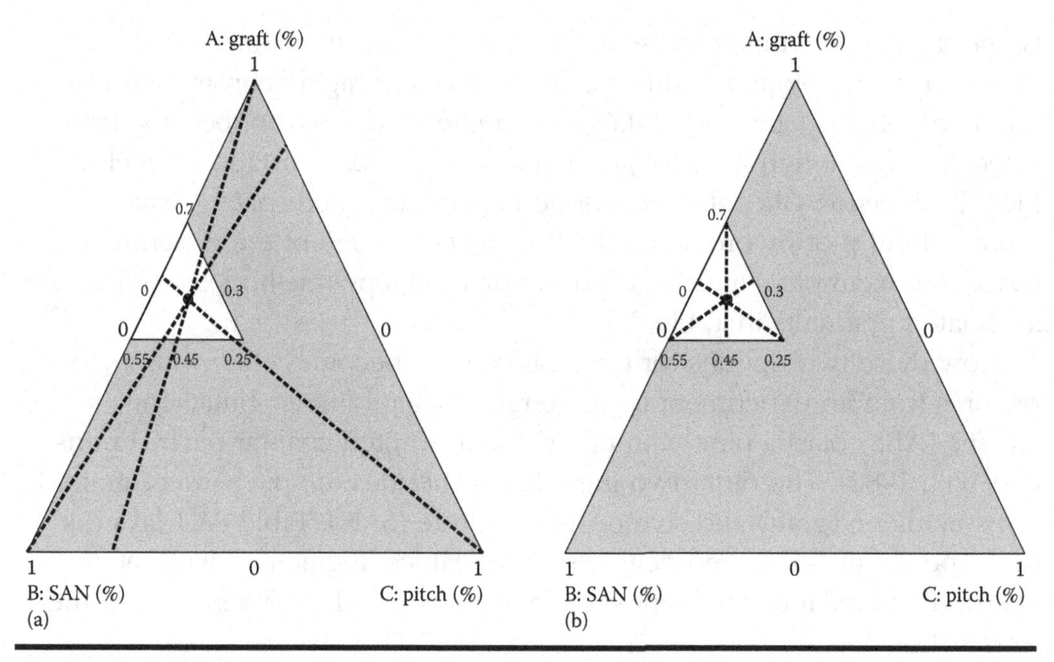

Figure 8A.1 Directions of trace for Cox (a) versus Piepel (b) in pipe case.

corners of the big triangle (coded in reals), whereas Piepel steers for the vertices of the small triangle (coded in pseudos').

The resulting trace plots are displayed in Figure 8A.2a and b.

The advantage provided by Piepel now becomes apparent by its tracks being of a consistent length and, for the most part, longer than Cox. Do not let the change in slope for component B from negative (Cox) to positive

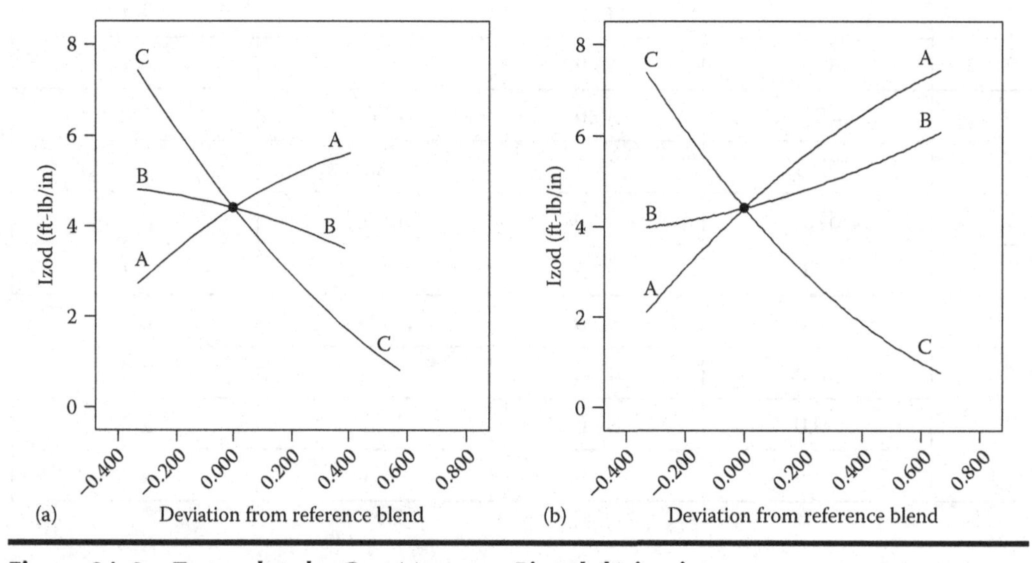

Figure 8A.2 Trace plots by Cox (a) versus Piepel (b) in pipe case.

(Piepel) put you off—this wavering indicates a linear effect that may not be significant and, indeed, pales in comparison to the impacts of A and C, which, throughout, remain consistent.

Consider that the traces are one-dimensional only, and thus, cannot provide a beneficial view of a response surface, especially with a nonlinear blending model. Furthermore, they depend not only on direction but also on the point of origin.

Chapter 9

Working Amounts, Categorical and Process Factors into the Mix

> The universe is like a safe to which there is a combination. But the combination is locked up in the safe.
>
> **—DeVries (1965)**

This chapter lays out designs that combine varying compositions of mixtures with changes in levels of process factors and other variables. The frosting on the cake, perhaps literally, is to experiment on two mixtures simultaneously, that is, a "mix–mix." These combined designs unlock a universe of potential synergisms. However, they can quickly expand far more runs than can be afforded. Therefore, they must be used judiciously.

Mixture-Amount Experiments—Not Just the Composition but How Much of It

We begin with the simplest combined experiment—one that incorporates a single process factor. A special case of this, first described by Piepel and Cornell (Piepel and Cornell, 1987), is a "mixture-amount" design (MAD).

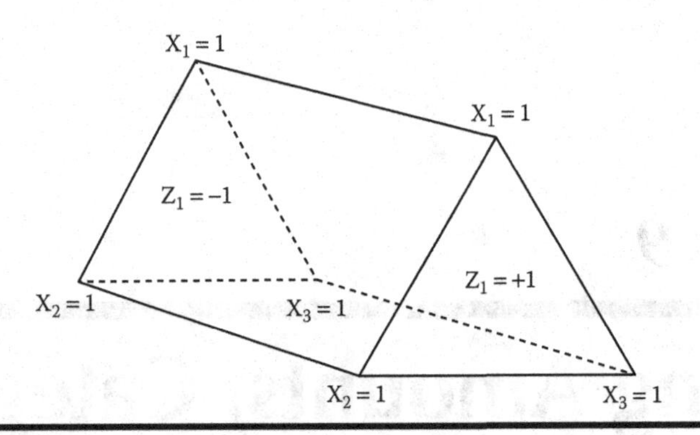

Figure 9.1 Three-component mixture-amount design.

A standard mixture experiment assumes that the response depends only on the proportions of the components present, for example, when tasting the food, a nibble will do. However, for such things as the effectiveness of fertilizer, the amount matters, as anyone with a lawn realizes after their grass comes out striped from variations in application. How much to use usually depends on the composition and vice versa—these two variables often interact.

Figure 9.1 depicts a combined design with three components and its amount (or one process factor) at two levels. Geometrically this three-component MAD forms a triangular prism. The mixture components define a (q−1) dimensional simplex: X_1, X_2,..., X_q. The amount contributes to additional unconstrained dimension: Z_1. The combined region of interest has (q−1 + 1) dimension—three in this case (3−1 + 1).

We will illustrate the structure of MAD and modeling of results via a case study on a controlled-release ibuprofen tablet (adapted from a study by Singh et al., 1995). The goal was to deliver just the right amount of this pain reliever over a period of 12 hours. The release properties depend both on the composition of the coating and its amount. Too little of a "weak" coating releases ibuprofen more quickly than desired. Conversely, an excessive amount of a "strong" coating falls short on immediate pain relief—measured at the first hour. The coating comprises three ingredients that total to 31% by weight/volume (w/v) in grams per milliliter (g/mL) of the solution:

1. Ethyl acrylate (EA) copolymer: 7–20 w/v%
2. Methyl methacrylate (MMA) copolymer: 7–20 w/v%
3. Triethyl Citrate (TEC) plasticizer: 4–14 w/v%

The thickness of the film is controlled by volume—the final factor in the combined design:

4. Coating amount, 15–25 mL

RELIEF OF THE PAIN OF PHARMACHEMICAL JARGON

Our mixture-amount case study involves poly(methyl) methacrylates (PMMAs), which are glassy copolymers that serve well as films for controlling the release of drugs. The polymerization occurs in an aqueous medium laden with surfactants to make the ingredients miscible. It all must be mixed very intensely to achieve emulsification (Roudsari et al., 2014). The result is a stable dispersion called "latex"—a term you may be familiar with in the context of water-based paints.

Incomprehensible jargon is the hallmark of a profession.

—Kingman Brewster, Jr. (Address to the British Institute of Management, December 13, 1977)

Two levels of amounts, as depicted in Figure 9.1, will be simplistic in this case—the rate of release is likely nonlinear as a function of coating thickness. Thus, the chemists decided to allow the amount to vary over three or more levels on a continuous scale, as dictated by an I-optimal selection based on the quadratic model:

$$Y(z) = \alpha_0 + \alpha_1 z_1 + \alpha_{11} z_1^2$$

This three-term model, when crossed with a quadratic mixture polynomial with six coefficients, creates an equation with 18 terms, which requires a like number of model points. Adding five checkpoints for testing lack of fit (chosen to fill space) and five replicates (selected optimally) brings the total of the MAD to 28 runs. The release results showing cumulative drug dissolved ("Dis") at two-time intervals are shown in Table 9.1. Underlined runs are replicates—these, along with all the others, being done in a randomized order.

Table 9.1 Results of Experiment on the Controlled Release of a Drug

ID	*A: EA* (*w/v%*)	*B: MMA* (*w/v%*)	*C: TEC* (*w/v%*)	*D: Coat* (*ml*)	*Dis 1 hr* (*cum%*)	*Dis 12 hr* (*cum%*)
1	20.0	7.0	4.0	15.0	59.2	93.5
2	13.5	13.5	4.0	15.0	37.9	99.3
3	7.0	20.0	4.0	15.0	64.8	88.3
4	15.0	7.0	9.0	15.0	77.9	100.0
5	7.0	15.0	9.0	15.0	76.9	80.9
6	10.0	7.0	14.0	15.0	82.6	100.0
7	10.0	7.0	14.0	15.0	75.4	91.7
8	20.0	7.0	4.0	17.5	42.0	77.3
9	7.0	20.0	4.0	17.5	57.0	74.0
10	11.0	11.0	9.0	17.5	51.6	69.6
11	20.0	7.0	4.0	20.0	40.2	74.3
12	13.5	13.5	4.0	20.0	9.8	87.3
13	7.0	20.0	4.0	20.0	39.1	57.0
14	15.0	7.0	9.0	20.0	41.1	79.9
15	7.0	15.0	9.0	20.0	47.3	64.0
16	10.0	7.0	14.0	20.0	49.5	76.7
17	10.0	7.0	14.0	20.0	53.7	69.9
18	20.0	7.0	4.0	22.5	31.2	54.6
19	11.0	11.0	9.0	22.5	32.6	54.1
20	20.0	7.0	4.0	25.0	25.6	54.0
21	13.5	13.5	4.0	25.0	4.0	55.1
22	13.5	13.5	4.0	25.0	3.1	54.1
23	7.0	20.0	4.0	25.0	1.1	48.0
24	7.0	20.0	4.0	25.0	1.9	40.7
25	15.0	7.0	9.0	25.0	14.0	42.9
26	7.0	15.0	9.0	25.0	15.5	20.2
27	10.0	7.0	14.0	25.0	13.3	42.1
28	10.0	7.0	14.0	25.0	27.2	43.8

Due to the size of the crossed models for combined designs such as this, we recommend that they be reduced. One of the better ways to eliminate terms is to take out any that fall below a specified p-value (typically 0.1) but do so in step-wise fashion, starting with the least significant one and going on from there until only the vital ones remain, that is, when no further reduction could be made. As noted in Chapter 7 for the tableting case, this is called the "backward" selection method (as opposed to "forward"—a method starting from a core model and adding the most significant term, next most, and so on). The two reduced models are shown below—all terms being significant but for AC in Y1, which comes in due to it being needed to support the hierarchy of ACD. (We detailed hierarchy in Appendix 1B. However, you may not have realized its repercussions on third-order terms such as ACD, which require not only the three main components (A, C, and D) be modeled, but also the three nonlinear blending terms: AC, AD, and CD.)

Y1. Dis 1 hr = 39.62 A + 37.01 B + 57.66 °C
Y2. − 78.63 AB − 10.66 AC − 15.98 AD − 33.93 BD − 21.48 CD
Y3. + 35.76 ABD − 53.01 ACD
Y4. Dis 12 hr = 71.52 A + 64.23 B + 70.92 °C
Y5. + 42.97 AB − 21.49 AD − 56.23 BC − 22.84 BD − 29.94 CD

The lack of fit on both models is insignificant ($p > 0.1$).

HOW LESS PROVIDES MORE FOR PREDICTED R-SQUARED

We promised, in the Chapter 1 sidebar "Why R-Squared Needs to Be Adjusted" that we would discuss using model reduction if the adjusted R-squared (R^2_{adj}) differs appreciably from the predicted R-squared (R^2_{pred}). Table 9.2 spells out the benefits of doing so in the case of the experiment on controlled release of ibuprofen. Notice how much the R^2_{pred} improves, especially on the second response (Y_2).

Other model statistics benefit from the model reduction but less dramatically. One exception is the raw R-squared, for which the larger models win, as always. That is simply wrong, as the case makes clear; thus,

(Continued)

Table 9.2 Statistics on Full versus Reduced Models

Statistic	Y_1 *Full*	Y_1 *Reduced*	Y_2 *Full*	Y_2 *Reduced*
Terms	18	10	18	8
Overall F	35.45	70.75	13.83	44.10
LOF p-value	0.4800	0.5754	0.1608	0.0582
R^2	0.9837	0.9725	0.9592	0.9391
R^2_{adj}	0.9559	0.9588	0.8898	0.9179
R^2_{pred}	0.4736	0.9236	−1.7614	0.8861

reinforcing that this unadjusted form of coefficient of determination cannot be relied upon for model selection.

Generally, we advise that when the R^2_{pred} falls to 0.2 or more units below R^2_{adj}, you try a model reduction. If that does not help, then consider a response transformation such as taking and converting them into logarithm scale (see, for example, Problem 5.1 in the 2nd edition of *RSM Simplified*). Another possible cause for such a discrepancy on the R-squared is a substantial block effect, in which case it cannot be rectified.

PS: A negative R-squared would normally be impossible, but in the predicted form, it can happen, for example, the R^2_{pred} for Y_2. This happens when the predicted residual sum of squares (PRESS) exceeds the residuals for the mean model, that is, less variation occurs around the average of the response than when it is fitted to the input variables. That is not acceptable. However, in this case, taking out the insignificant terms brings the R^2_{pred} well into positive territory.

Being significant terms for both responses that include factor D indicates that amount does affect the release of the drug as one would assume. The pharma chemists hoped to achieve 10%–30% dissolution at one hour, for the same composition and amount of coating, a release of 80%–100% at 12 hours. They succeeded as seen by the window that appears in the graphical overlay depicted in Figure 9.2.

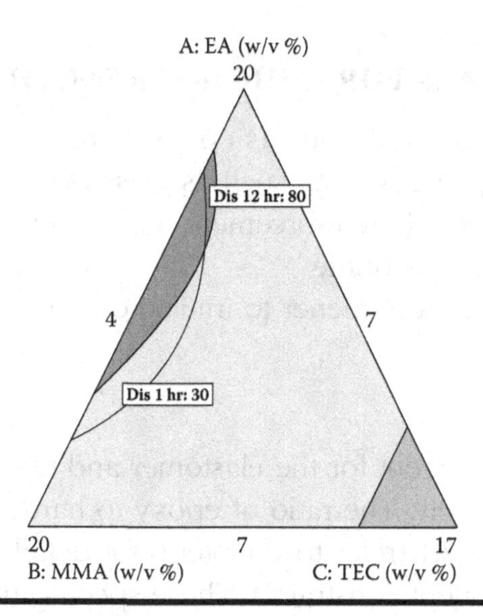

Figure 9.2 Sweet spot for the release of ibuprofen at coating amount (D) of 19 mL.

This elusive sweet spot occurs only between 16.5 and 19.7 mL of the coating and only for a particular combination of ingredients. Only by a well-conceived combined DOE can such desirable outcomes be found in a reasonable number of runs.

Contending with Categorical Variables

Categorical variables come up frequently in industry, for example, when a purchasing agent seeks competitive vendors, or a chemist wants to try substitute materials. An adjustment to the mixture recipe may be required to facilitate these alternatives. To illustrate how to set up such a combined design, let's look at a hypothetical case on the material science of a composite material used in aerospace. The mission is to:

1. Select one of two elastomers.
2. Choose a fiber from three alternatives.
3. Optimize the recipe of elastomer, fiber, hardener and epoxy resin.

Elastomer and fiber types are categoric factors; that is, only one of each can be present in any given mixture.

THE ESSENTIALS FOR A HIGH-STRENGTH COMPOSITE

In composites, one material, such as epoxy resin, serves as a "matrix" to hold everything together. Fibers, such as glass or carbon, when embedded in the matrix provide reinforcement. Elastomers are added to make the whole composite less brittle.

PS: Epoxies require a hardener to initiate curing.

The ranges in weight percent for the elastomer and fiber are 5%–10% and 54%–62%, respectively. The ratio of epoxy to hardener must be maintained from 1.8 to 2.1 (refer to Chapter 6 for detail on how to convert this into a multicomponent constraint). These specifications set the ranges for hardener and epoxy levels somewhere between zero and one-hundred percent (0%–100%)–an algorithm such as that detailed in Appendix 5A will get this job done. Nothing about the mixture part of the combined design is new at this stage in the book—the only aspect of it is being combined with categorical variables. Figure 9.3 shows how this complicates the structure.

The minimal number of points for an optimal design expands exponentially as more categorical variables are added, especially, when they expand beyond two types each, as in the case of the fiber alternatives. To keep the design to a manageable level, this experiment on composites is set up for a quadratic mixture model with 10 terms (the 4 main components plus 6 second-order nonlinear blending combinations) crossed by a two-factor interaction equation for the

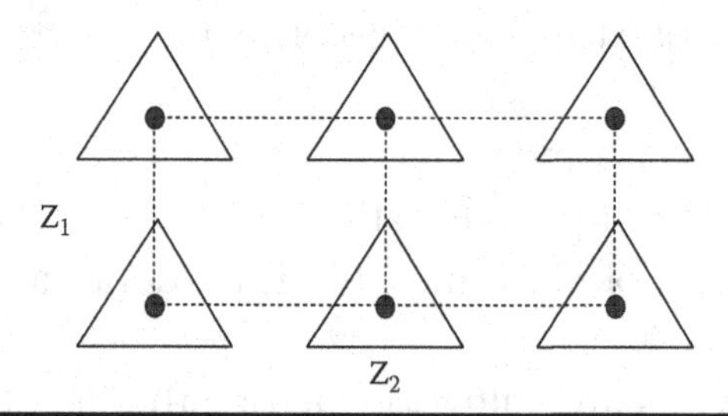

Figure 9.3 Four-component mixture combined with two categorical variables.

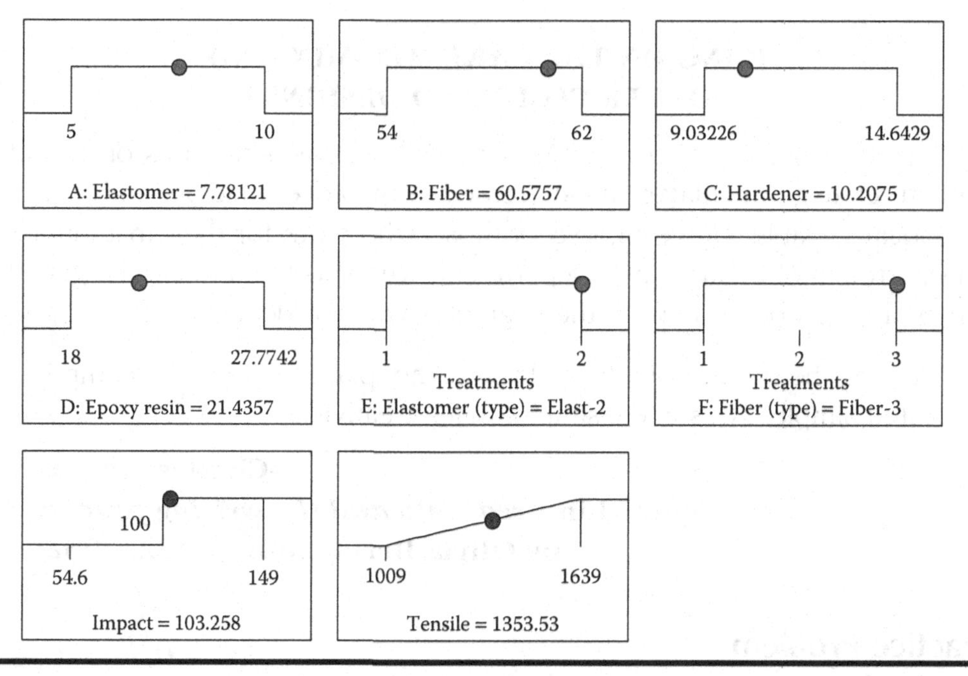

Figure 9.4 An optimal composite recipe made with selected types of elastomer and fiber.

categorical variables (Z_1 and Z_2) with 6 coefficients (c_0, c_1Z_1, $c_{2[1]}Z_2$, $c_{2[2]}Z_2$, $c_{12[1]}Z_1Z_2$, $c_{12[2]}Z_1Z_2$), where the square-bracketed numbers deal with the two contrasts between the three types of fiber. This is complicated! The design then must provide a base of 60 model points (10 mixture times 6 categorical). Adding 5 checkpoints for lack of fit and replicating 5 combinations for pure error brings the design to 70 points in total.

This example is made up, but to provide an idea of where it might lead, we offer Figure 9.4 to spell out the feasible ranges for the hardener and epoxy (these were allowed to float initially), and a hypothetical solution where impact strength of the composite must be at least 100 kilojoules per meter squared and tensile properties maximized.

This material-science example only scratches the surface (pun intended) of what can be done and how much work will be required when combining mixtures with categorical variables. Before going down this road, consider simplifying things by first sorting out which materials to use, whom to get them from and so forth—using a standard recipe, and then reoptimize the formulation once you settle on the categorical variables.

ICING ON THE CAKE MIX-MIX AND OTHER COMBINED DESIGNS

We could continue with many other combinations—mixtures on top of mixtures (e.g., multilayer films) crossed with process factors and/or categorical variable. However, except those who work for their institution's Department of Redundancy Department, we expect you to get the gist of how this can be done from the examples we provided.

> Making the simple complicated is commonplace; making the complicated simple, awesomely simple, that's creativity.
>
> **—Charles Mingus**
> **(quoted in *Creativity and the writing process* by Olivia Bertagnolli, p. 182, 1982)**

Practice Problem

To practice using the statistical techniques you learned in Chapter 9, work through the following problem.

Problem 9.1

Experience how easy it will be to design a mixture experiment, combined with multiple factors (one of the combinations we skipped over) and analyze the results by using a computer tool specialized for this purpose. It can be freely accessed via the website developed in support of this book: See *About the Software* for the path. When you arrive at the internet page, follow the link to the accompanying tutorials. Then download and print the one on *Combined Mixture-Process*. It explains how food scientists came up with just the right texture (never mind the taste!) for fish sticks made from an optimal blend of mullet, sheepshead, and croaker (yuk!). They were then deep-fried for an ideal period before being baked to perfection at the ideal temperature and time settings. You will be amazed.

Appendix 9A: Alternatives for Modeling Results from Combined Designs

The crossed models we detailed in this chapter are straightforward to create but, as the number of components and factors increase, they become very large with many terms going beyond second order that end up being taken out due to their insignificance. Chapter 11 in *RSM Simplified* provides one good alternative—model the mixture components in terms of ratios and simply incorporate them into factorial or RSM designs alongside the process factors. However, the mathematics required by the ratio approach make this impractical. An equally efficient (for a number of model terms), but less intense approach, is to selectively remove terms from the crossed model, as advised by Kowalski, Cornell, and Vining (Kowalski et al., 2000). These "KCV" models are geared to a second-order approximation via an appealing "intuitive balance" (p. 2279). They preserved the view of mixture-process interactions that experimenters are seeking when they deploy combined designs. For these reasons, we feel that KCV models deserve detailing.

The advantage of KCV models comes from adding, rather than crossing, higher order terms from the equations for the mixture and process variables. To illustrate the difference, let's apply the KCV model to the fish-patty experiment of Problem 9.1, which involves three mixture components ($q = 3$) and three process factors ($k = 3$). The crossed quadratic-by-quadratic (QxQ) model requires 60 terms from the multiplication of the six in the mixture (X_1, X_2, X_3, X_1X_2, X_1X_3, X_2X_3) and ten for the process (Intercept, Z_1, Z_2, Z_3, Z_1Z_2, Z_1Z_3, Z_2Z_3, Z_1^2, Z_2^2, Z_3^2). It includes 18 fourth order terms (e.g., $X_1X_2Z_1Z_2$), which can be safely removed by capping the crossed model at cubic. This reduces the model to 42 terms. The quadratic KCV model (and ratio RSM) requires only half of the reduced crossed QxQ—21 terms (X_1, X_2, X_3, X_1X_2, X_1X_3, X_2X_3, X_1Z_1, X_1Z_2, X_1Z_3, X_2Z_1, X_2Z_2, X_2Z_3, X_3Z_1, X_3Z_2, X_3Z_3, Z_1Z_2, Z_1Z_3, Z_2Z_3, Z_1^2, Z_2^2, Z_3^2). As shown in the following general equation, it is generated by crossing the linear models and then adding the second-order terms:

$$\eta(x,z) = \sum_{i=1}^{q} \beta_i x_i + \sum_{i<}\sum_{j}^{q} \beta_{ij} x_i x_j + \sum_{i=1}^{q}\sum_{n=1}^{k} \gamma_{ik} x_i z_n + \sum_{n<}\sum_{m}^{k} \alpha_{nm} z_n z_m + \sum_{n=1}^{k} \alpha_{nn} z_n^2$$

where q and k are the indices for the process factors and the mixture components, respectively. The interactions between the linear blending terms in the mixture and main-effect process variables (e.g., X_1Z_1) provide the core value for the crossing of models, that is, an opportunity to find just the right combination of component levels at a particular set of factors.

Table 9A.1 Number of Terms in Various Models for Combined Mixture-Process Experiments

Mixture Components	*Process Factors*	*Crossed QxQ All*	*QxQ Max Cubic*	*Ratios (#) RSM Model*	*KCV*
2	2	18	15	10 (1)	10
3	2	36	27	15 (2)	15
4	2	60	42	21 (3)	21
3	3	60	42	21 (2)	21
4	3	100	64	28 (3)	28
5	3	150	90	36 (4)	36
4	4	150	90	36 (3)	36
5	4	225	125	45 (4)	45
6	4	315	165	55 (5)	55

Table 9A.1 shows, for increasing numbers of mixture components (2–6) and process factors (2–4), how KCV (hand-in-hand with the ratio approach) fares far better than the crossed QxQ model for combined designs, even when the terms are kept to a maximum of cubic (3rd order). As the designs build beyond the underlined combination for the fish-patty case (q = 3, k = 3), the practical advantage of KCV becomes pronounced.

NONLINEAR COMBINED MIXTURE MODELS

Ronald D. Snee, a pioneer in the field of mixture design, suggests using nonlinear estimation techniques to fit models that are "nonlinear in the parameters" (private correspondence to MJA 6/10/15), for example, exponential [$Y = f(X^k)$]. However, application of nonlinear modeling to combined mixture-process experiments (Snee et al., 2016) requires users to specify the equation that relates the coefficients to the response to understand and use, without providing any advantage for reducing runs, but this strategy is more complicated than KCV.

Chapter 10

Blocking and Splitting Designs for Ease of Experimentation

The generation of random numbers is too important to be left to chance.

—Robert R. Coveyou, title of article in *Studies in Applied Mathematics*, III (1970)

Randomization protects experimenters against confounding caused by lurking variables such as rising ambient temperatures run-by-run. However, it pays to restrict randomization under the following circumstances:

- Limited availability of resources such as material, time, equipment or operators, which then must be carefully managed so the variability caused by shifting them over in mid-experiment does not create bias in effect estimates.
- Input variables being very hard to change (HTC).

These limitations lead to groupings of runs via the block and split plots, respectively.

Blocking to Remove Known Sources of Variation

We introduced blocking in Chapter 2 for the beer-blending case where it filtered out differences between the three tasters. This experiment being fully replicated by taster presented no difficulty in design. Blocking becomes far more problematic when it involves a breakdown of a given mixture design into subgroups. If it is done without careful thought, model parameters can become highly correlated with the block effect. To avoid such biasing, Draper et al. and Lewis et al. developed templates for orthogonal blocking of simplex designs for a limited number of components (Draper et al., 1993) that they then generalized (Lewis et al., 1994). We will lay out one of these Draper–Lewis designs to provide a picture of well-blocked simplex. This sets the stage for a more-flexible method of blocking mixture experiments—applicable to simplex or nonsimplex regions—which makes use of modern optimal-design tools.

LATIN SQUARES FRAME OUT BLOCKS ORTHOGONALLY

The Draper–Lewis blocking schemes use Latin squares as their basis. These are square layouts with Latin letters with a number of rows and columns equal to the number of treatment levels. They must be constructed such that each treatment occurs only once in each row and once in each column. For example, see Table 10.3 (p. 218) in *RSM Simplified, 2nd edition*, which lays out a balanced plan to test 4 tires being rotated around the 4 positions in a car.

For more details on constructing orthogonal blocks in mixture designs using Latin squares see Chapter 8.4 (p. 447) of *Experiments with Mixtures* (3rd ed., Wiley, 2002) by John Cornell.

The blocked mixture experiment cited by Draper et al. and Lewis et al. in their 1993 and 1994 publications, respectively, was conducted by an English miller on four flours, each derived from a different variety of wheat. The food chemists blended these four components into doughs in various proportions, which, when baked into bread, created the varying results (simplified somewhat) in specific volumes shown in Table 10.1. The higher the volume the better for it to produce loaves of least density, as desired by consumers in Great Britain. By breaking this experiment into two blocks, it lessened the impact of unknown time-related variables that might otherwise be confounding. The block size of 9 runs, each fell well within the capacity of the pilot-scale bakery.

After removing the variation due to the blocks, the regression produces a model that exhibits strong nonlinear blending of flours 1–3 with flour 1,

Table 10.1 Blocked Mixture Experiment on Bread-Flours

Run	*Block*	*A: Flour 1*	*B: Flour 2*	*C: Flour 3*	*D: Flour 4*	*Volume (ml/100 g)*
1	1	0	0.25	0	0.75	392.0
2	1	0.25	0	0.75	0	423.5
3	1	0	0.75	0	0.25	427.0
4	1	0.75	0	0.25	0	423.0
5	1	0	0.75	0.25	0	421.5
6	1	0.25	0	0	0.75	423.5
7	1	0	0	0.75	0.25	378.0
8	1	0.75	0.25	0	0	426.5
9	1	0.25	0.25	0.25	0.25	413.0
10	2	0	0.25	0	0.75	389.5
11	2	0.25	0	0.75	0	421.0
12	2	0	0.75	0	0.25	413.5
13	2	0.75	0	0.25	0	416.5
14	2	0	0	0.25	0.75	350.0
15	2	0.25	0.75	0	0	431.0
16	2	0	0.25	0.75	0	384.5
17	2	0.75	0	0	0.25	408.5
18	2	0.25	0.25	0.25	0.25	429.0

that is, terms AB, AC, and AD. The strong influence of this first flour (A) can be seen in the trace plot shown in Figure 10.1.

The mean square for the blocks exceeds the residual mean square by nearly fourfold, which is appreciable, that is, only by filtering out this variation over time was it possible for this design to reveal the formulation behavior exhibited in the trace plot.

This experiment on bread flour pioneered the application of blocking to mixture designs. However, modern-day computational tools (such those provided by the software accompanying this book) provide more flexible (not restricted to simplex regions) and better (statistically optimal) designs for any number of blocks within reason. Applying optimal design, in

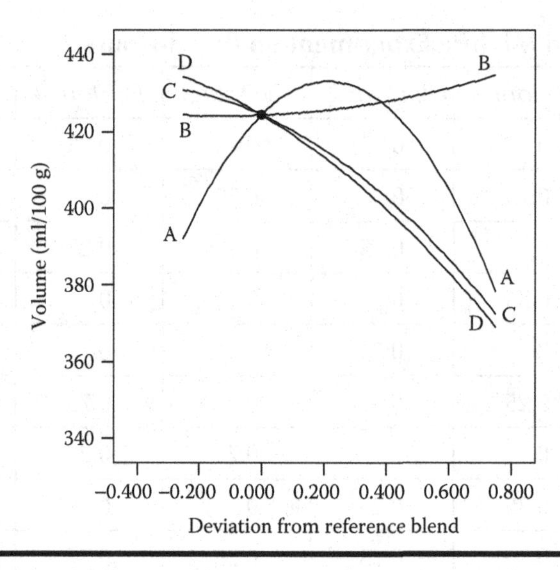

Figure 10.1 Trace plot of effects of flours on baked volume of bread.

retrospect to the flour-blending case, we produced the experiment shown in Table 10.2. It incorporates four pure replicates, which enables a test lack of fit—a shortcoming of the original design by Draper and Lewis. This design improvement comes at a small price—one extra run per block.

THE POSSIBILITY OF THIRD-ORDER NONLINEAR BLENDING

As you may have noticed, beyond the overall centroid, the original flour-blending design restricts the blending to only two components at a time, thus it cannot estimate the third-order special cubic term in the mixture model. Also, as we've already mentioned, this experiment lacked the necessary elements (check blends plus replicates) to provide a test for lack of fit. Given the strong nonlinear blending impact of flour #1 (A) with all other flours, perhaps this might be better modeled by 3rd order special cubic terms (ABC, ACD, ABD). We will never know.

The rebuilt design includes numerous blends of three or more components and it does support the fitting of all special cubic terms. It also tests lack of fit beyond that. Thank goodness for advancements in programming along with the leaps in computing power since 1993 when blocked

(Continued)

mixture designs were still in their infancy. To be fair, all of these, coming at a very slight cost-optimal designs, such as those we laid out, are not orthogonally blocked, only nearly so (e.g., VIF of 1.4 in our design—a value of 1.0 being completely orthogonal).

> Orthogonally blocked mixture experiments [such as the bread-baking one] are highly inefficient compared to optimal designs.
>
> **—Goos and Donev (2003)**

Table 10.2 Flour Experiment Rebuilt with Modern Tools for Optimal Design

ID	*Block*	*Build Type*	*Space Type*	*A: Flour 1*	*B: Flour 2*	*C: Flour 3*	*D: Flour 4*
1a	1	Model	CentEdge	0.500	0.500	0.000	0.000
1b	1	Replicate	CentEdge	0.500	0.500	0.000	0.000
2	1	Model	Vertex	0.000	1.000	0.000	0.000
3a	1	Model	CentEdge	0.500	0.000	0.500	0.000
3b	1	Replicate	CentEdge	0.500	0.000	0.500	0.000
4	1	Model	CentEdge	0.000	0.500	0.500	0.000
5	1	Model	Vertex	0.000	0.000	1.000	0.000
6	1	Model	PlaneCent	0.000	$0.3\overline{33}$	$0.3\overline{33}$	$0.3\overline{33}$
7	1	Model	Interior	0.375	0.125	0.125	0.375
8	1	Lack of fit	AxialCB	0.125	0.125	0.125	0.625
9	2	Model	Vertex	1.000	0.000	0.000	0.000
10	2	Lack of fit	PlaneCent	$0.3\overline{33}$	$0.3\overline{33}$	$0.3\overline{33}$	0.000
11	2	Lack of fit	AxialCB	0.125	0.625	0.125	0.125
12	2	Lack of fit	AxialCB	0.125	0.125	0.625	0.125
13a	2	Model	CentEdge	0.500	0.000	0.000	0.500
13b	2	Replicate	CentEdge	0.500	0.000	0.000	0.500
14a	2	Model	CentEdge	0.000	0.500	0.000	0.500
14b	2	Replicate	CentEdge	0.000	0.500	0.000	0.500
15	2	Model	CentEdge	0.000	0.000	0.500	0.500
16	2	Model	Vertex	0.000	0.000	0.000	1.000

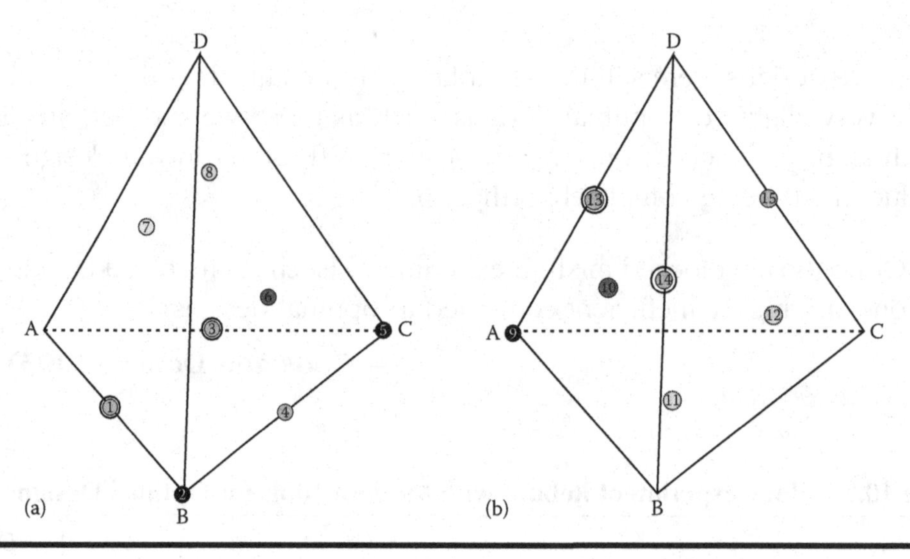

Figure 10.2 Point locations in blocks 1 (a) and 2 (b) for optimal design on flours.

Figure 10.2 shows the optimal design block by block within the tetrahedral space. The double-circled points are replicated.

Note how the point allocation in our optimal design is mutually exclusive, that is, they complement each other by block, whereas in the original experimenters reran blends 1–4 and 9 in their second block. If you do want a "control" blend across both blocks, we recommend it be the overall centroid. Also, since three out of the four vertices made it into this algorithmic design, adding the missing one (D—the pure blend of flour 4) would be sensible.

To summarize, using standard templates, or more-versatile computer tools for optimal design, blocks should be set up to be orthogonal, or nearly orthogonal, to the component effects you want to estimate. Having done so, the breakdown of runs into subgroups will not make much of a change to the model-coefficient estimates, if at all. Then if blocks differ in an appreciable degree, you need not worry, because this variation will be filtered out of the overall system noise, thus making it easier to detect significant changes to the response caused by the components themselves.

Split Plots to Handle Hard-to-Change Factors or Components in Combined Designs

To satisfy statistical assumptions, properly executed experiments require runs to be completely independent. This can be very daunting for researchers running combined designs such as those detailed in Chapter 9, because:

- A new blend must be prepared even if the recipe listed in the experimental plan does not change.
- The process must be reset, even though all of the factors remain at the same level in the design from one run to the next.

If all the variables are easy to change (ETC), this complete randomization with resets should be done. However, it is not often practical, or even possible, to perform an experiment in this way. Consider, for example, a food scientist mixing up a single cookie by the recipe specified in the design layout and then baking it all by itself per the plan for that run. That would be ridiculous. A more sensible approach in this case would be to mix a batch of cookies with varying recipes and then bake a tray of them. This introduces a restriction in randomization that creates a designed-experiment called a "split plot."

APPLYING A SPLIT PLOT TO MIXTURE ONLY

The big advantage of split plots comes with combined experiments as we've discussed. However, it is conceivable that some components might be harder to change than others, for example, in a heterogeneous reaction involving a liquid reagent catalyzed by a mixture of solid ingredients. In this case, it would be sensible to create master batches of the liquid components and then add various blends of the solid catalysts. This would be a mix-by-mix split plot.

A silicone base (Mixture 1-HTC) is prepared and split up, after which a different catalyst package (Mixture 2-ETC) is added to each piece.

For background on split plots see Chapter 11 of *DOE Simplified, 3rd edition.* Here we will detail its application to an experiment done at Stat-Ease headquarters to improve the office coffee by a better blend of beans, combined with improved methods for grinding and brewing (Bezener and Anderson, 2016).

The coffee provided by the vendor of the Stat-Ease brewing machine created a backlash from our programmers who declared it to be "disgusting and unacceptable." They questioned not only the type of coffee being used but also the fineness of its grind and how much of it to be brewed per pot. The team agreed that changes would be made only if the new coffee be judged better on average by a core group of 5 drinkers, and no one should hate it. This second requirement essentially gave each of the individual testers the power to veto the majority's rule.

The experiment combined a mixture-amount design (discussed in Chapter 9) with a categorical factor—the grind size. The input variables were:

- Components a, b, and c*: Light, Medium, and Dark roast *(lower-case lettered to denote them being HTC and thus grouped by blend)
- Amount (factor D): 2.5–4.0 ounces of beans (continuous)
- Grind (E): Fine, Medium, or Coarse (categorical)

The fully randomized experiment required freshly blended beans for each pot. However, it being far more convenient to mix these up in quantity, the blends were restricted to 16 groups using a 74-run split-plot design. Table 10.3 shows the first 9 runs of the experiment to illustrate the grouping (3 shown) of the mixtures.

Table 10.3 First 9 Runs of Coffee Experiment

Group	*Run*	*A: Light*	*B: Medium*	*C: Dark*	*D: Amount* (oz.)	*E: Grind*	*Avg Liking:* (1–9:)	*Min Liking:* (1–9:)
1	1	0	0	1	4.00	Medium	6.4	5
1	2	0	0	1	2.50	Coarse	5.2	4
1	3	0	0	1	3.25	Fine	5.2	3
1	4	0	0	1	2.50	Medium	5.0	4
2	5	0.5	0.5	0	4.00	Fine	4.2	3
2	6	0.5	0.5	0	3.25	Medium	5.6	5
2	7	0.5	0.5	0	4.00	Coarse	5.4	4
2	8	0.5	0.5	0	2.50	Fine	5.2	3
3	9	0.5	0	0.5	3.18	Fine	5.4	4
~	~	~	~	~	~	~	~	~

Each blend of coffee beans was, on average, tested at 4 amount-grind size combinations. The barista (Martin) randomly interspersed six runs of the current office coffee throughout the experiment to serve as a control.

LIGHT COFFEE HEAVIEST FOR CAFFEINE?

It is natural to assume that dark roast has more caffeine than a lighter coffee. The true "caffiends" argue that roasting breaks down caffeine so they go for the light. For that reason breakfast blends are generally a lighter roast. However, both of those schools of thought may be wrong because all roasts of the same bean have basically the same amount of caffeine, at least according to one expert (Thomson and Thomson, 2016).

Another variable involves how you measure your coffee. If by scoops, light coffee will provide a bit more caffeine, because roasting drives out water and thus reduces the density of darker beans (Kicking Horse Coffee. Caffeine myths: Dark vs. light. www.kickinghorsecoffee.com/en/blog/caffeine-myths-dark-vs-light). We advise you weigh out your coffee and thus avoid this confounding variable.

Perhaps the best advice is to drink what you find tastiest. Then you will imbibe more and dose up on caffeine more heavily, as needed for whatever task you have at hand.

A mathematician is a device for turning coffee into theorems.

—Alfréd Rényi, Probability Theorist (Suzuki, 2002)

A supplement to the book details the modeling and statistical analysis of this combined split-plot design. A subsequent numerical search came up with a most desirable 50–50 blend of medium and dark beans (no light ones) when ground to the fine level and brewed at a loading of 2.5 oz. This hit the "sweet spot" for the taste testers as evidenced by 10 follow-up runs—4 of the chosen blend, 2 with the standard office coffee, and 4 at various other combinations of beans ground in different ways and produced with changing amounts, none of which deviated significantly from the model predictions.

A Case Where the Process, Rather Than the Mixture, Is Hard to Change

Kowalski et al. the inventors of the KCV model we touted in Appendix 9A, provide an example (Section 5 of Kowalski et al., 2002) of process factors being HTC in a design combined with ETC variations in mixture components. This contrasts with the coffee case where the blending was HTC. Figure 10.3 illustrates the primary options for a split-plot, with combined design either being mixture HTC—a triangle with squares at the vertices or process HTC—a square with triangles at the corners.

The KCV case, pictured by Figure 10.3b, was aimed at modeling the thickness of a vinyl car-seat covering as a function of the plasticizer composition and processing conditions. The mixture side of the experiment studied pure blends, binary blends, and the centroid of three plasticizers (P1, P2, and P3) randomized (not a problem being ETC) within groups. The process factors were extrusion rate and drying temperature set at two levels, each in 4 groups (these being HTC). Three more groups were run at the center point of the process factors with the centroid blend replicated 4 times. This is a good example of a split plot being put to good use for the practicality of processing.

In conclusion, split-plot designs provide accommodating layouts, especially when experimenters want to combine mixture components with process factors and one or the other is hard to change. Computer-aided optimal selection of runs facilitates customization of group sizes to make things even more convenient.

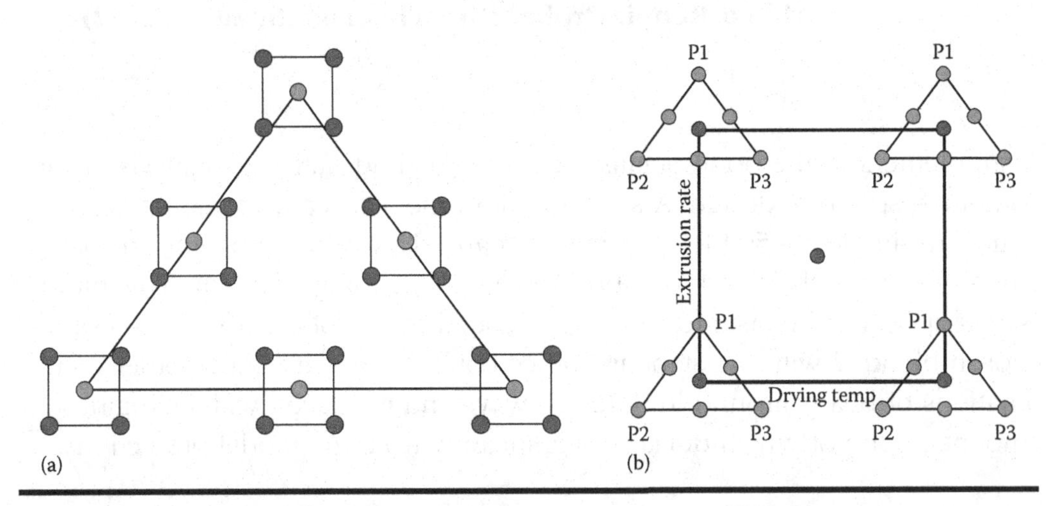

Figure 10.3 Split plots for mixture HTC (a), versus process HTC (b).

Chapter 11

Practical Magic for Making the Most of a Mixture

> Eye of newt and toe of frog, wool of bat and tongue of dog. Adder's fork and blindworm's sting...Barbados lime is just the thing. Cragged salt like a sailor's stubble! Flip the switch and let the cauldron bubble!
>
> **—Owen's sisters (witches) reciting recipe for margaritas (Practical Magic, Warner Brothers, 1998)**

In this last chapter, we come back to what defines a mixture and provide practical advice on how to put the statistical design of experiments to good use for optimizing your formulation. To begin, let's dispel various excuses to avoid the peculiarities of mixture design and analysis by denying the true nature of the experimental variables.

The Failure of Fillers and Perils of Parts

As we advised in Chapter 1, you'd best deploy the statistical tools for mixture design and analysis when you experiment on a formulation for which the response depends on proportions, not amounts, of ingredients. For this to work, you must fix the total of your components that will be varied, for example, a 20-gallon cauldron for a magical concoction. Unfortunately, many formulators, particularly those who are only taught the standard DOE tools for factorials and response surface methods, do not take to mixture design. They excuse themselves along the following two lines epitomized by these actual quotes from clients who must remain nameless:

1. Filler (ignoring the main ingredient): "The process engineers do not see a clear reason why their 2^3 factorial was inappropriate. They told me that the filler was "like the ocean" compared to the amounts of the other three components, so they didn't see it as a variable to have in the model."
2. Parts (not including the base as a component): "I am doing a chemical formulation DOE study. I have two ingredients to study in relation by parts to the base polymer (100 parts). So, I don't think it is a mixture design."

The first objection is easiest to address because "filler" is an oxymoron. These materials would not be included in the recipe if they did not perform some function if only to keep extremely potent "active" ingredients diluted. Beyond that, allowing the studied components to vary in total creates redundancies in how the ingredients relate to each other by proportion. To illustrate this latter issue, consider the following failed experiment brought to one of the authors (Pat) for a postmortem. It involved the study of a combination of chemicals called "dopants" that created various colors on a cathode ray tube (CRT), such as those used in televisions before flat panels took over. These were mixed in quantities of parts per billion into inert filler. Naturally, the experimenters figured they'd simply run a factorial design on the dopants. However, due to the dependency of color on proportions of these ingredients, this approach failed miserably. Consider mixing one blue and one yellow, versus two blues and two yellows—either way, the picture goes green. The ratios of 1:1 versus 2:2 create similar effects. Unfortunately, without employing the right tool for this problem—a mixture design—the experimenters unwittingly, are limited, relative to their exploration of possible proportions of incredibly potent components.

APPLYING FACTORIALS AND RESPONSE SURFACE METHODS (RSM) TO MIXTURES

We layout at the outset of Chapter 11 of *RSM Simplified, 2nd ed.* on "Applying RSM to Mixtures" how a factorial design fails for a smoothie comprised of apples and bananas. This experiment features both a 2-to-1 and a 4-to-2 ratio of these two fruits, which simply scales up the amount of beverage without changing its taste. Factorials and RSM can work well for cases like this, but only if the components are varied by ratios. For example, a one-factor experiment on the ratio of apple to banana ranging from 1 to 2 would be good. Beyond two components this approach becomes more complex, as we spell out later in Chapter 11. Using mixture design, when supported by the proper computing tools, is a lot easier than going the ratio route.

The second objection to mixture design, the one based on formulating by parts, merits an in-depth rebuttal because of this seductively simple methodology being so widespread.

Many years ago, we worked with a client who learned RSM. Naturally, they were very excited by the possibilities of this powerful statistical tool and applied it immediately to the following recipe for a silicone sealant:

1. Plasticizer varied from 50 to 100 parts
2. Filler (inorganic) varied from 100 to 250 parts
3. All other ingredients held at 57 parts
4. Polymer set at constant level of 100 parts

The adhesive chemists set up a 13-run central-composite design on the filler and plasticizer. Figure 11.1 shows the experiment in a standard order (left to right) with the four two-level (2^2) factorial core runs first (1–4), then the four axials (5–8), and finally, the five center points (9–13).

Unfortunately, as you now can plainly see by the differing bar-heights, the total amount going into the reactor changed with every new combination of ingredients. The proportions of the other ingredients (C) and polymer (D) also vary. Their relative concentrations range up and

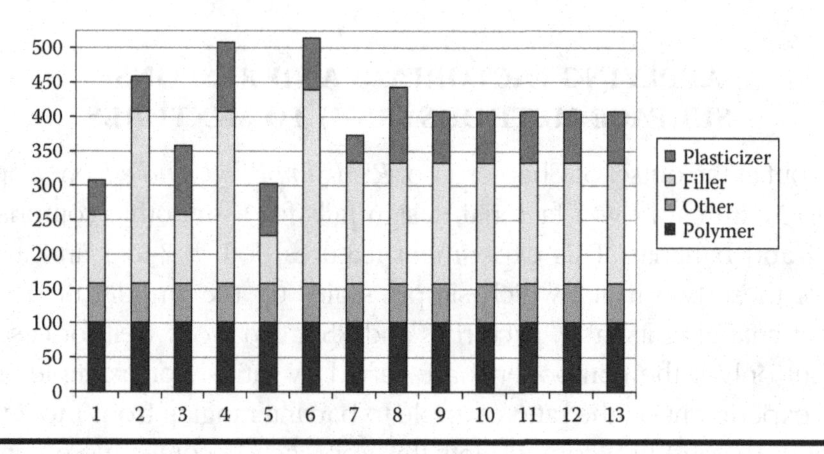

Figure 11.1 Sealant experiment laid out by parts.

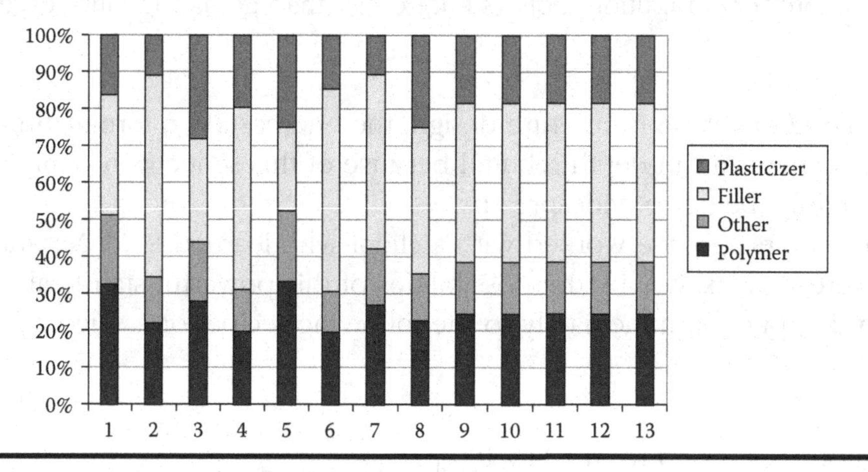

Figure 11.2 Sealant experiment laid out by percentages.

down in a very disorganized fashion as illustrated in Figure 11.2, which redraws the bar chart by percentages.

The ingredient ranges by weight percent were:

1. Filler, 23%–55%
2. Plasticizer, 11%–28%
3. Other, 11%–19% (not a constant!)
4. Polymer, 19%–33%

The factors studied were ingredients and performance as a function of proportions: This is a mixture! Fortunately, this persuasive presentation helped the formulators see the light before they wasted a lot of time. With our guidance, the chemists redesigned their experiment as a three-component mixture:

1. Filler 25%–55%
2. Plasticizer 11%–28%
3. Polymer 20%–33%

The remainder of 14% was held constant in a thirteen-blend design.

To wrap up this case of the perils of parts, check out the bar chart in Figure 11.3 displaying the final mixture design.

Notice how the bottom bar remains constant as originally intended, the chemist only wanting to vary the three components that make up 86% of the fixed total. This is a good case against the use of parts in a factorial or RSM design. It's best to use mixture designs to experiment on formulations.

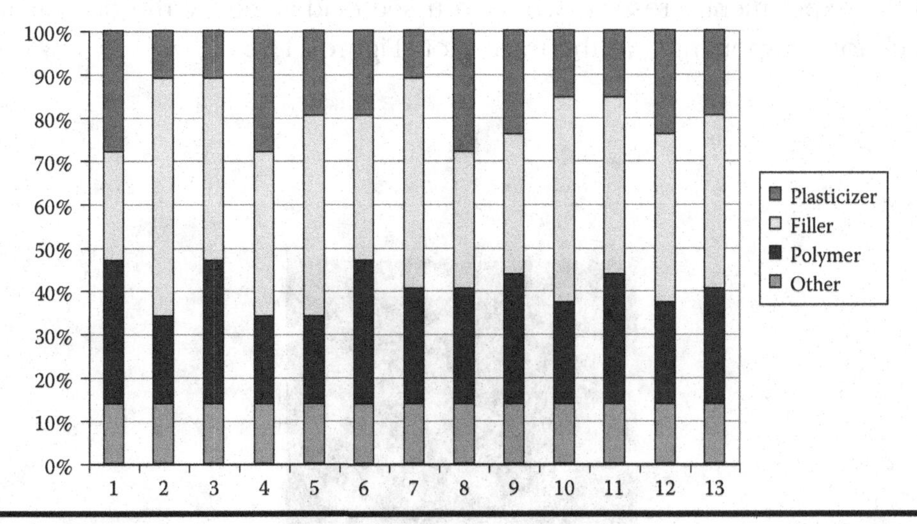

Figure 11.3 Sealant experiment done with a mixture design.

MIXTURES ARE LIKE A BOX OF CHOCOLATES

Hollywood's stereotype of a mixture experimenter is a mad scientist—typically a bug-eyed fellow with very thick glasses and a few wisps of hair badly combed over his balding pate—crowing over a smoking beaker of alarmingly colored chemicals. However, a box of chocolate-covered cherries is a mixture too, as recognized by an astute female food scientist far from the movie character described above. With our help (paid for in candy), she set up an experiment that modeled consumer response to various packing defects for their confection:

1. Upside down, 0%–33%
2. Sideways, 0%–33%
3. Leakers, 0%–33%
4. OK, the remainder to total 100%

The chocolatiers were most concerned by pieces that leaked (component C) when the sticky filling burst their bubble in transit as seen pictured in Figure 11.4.

This they knew would create complaints, and it turned out to be so in the experimental results, which you see evidenced by the downward track for component C in the trace plot (Figure 11.5).

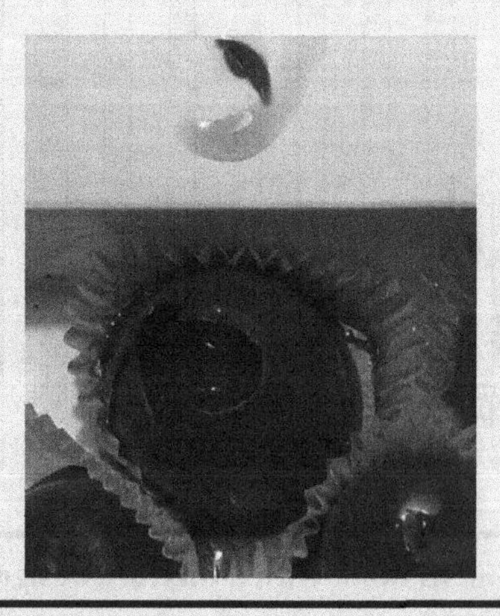

Figure 11.4 Cherry filling leaking out of a broken chocolate.

(Continued)

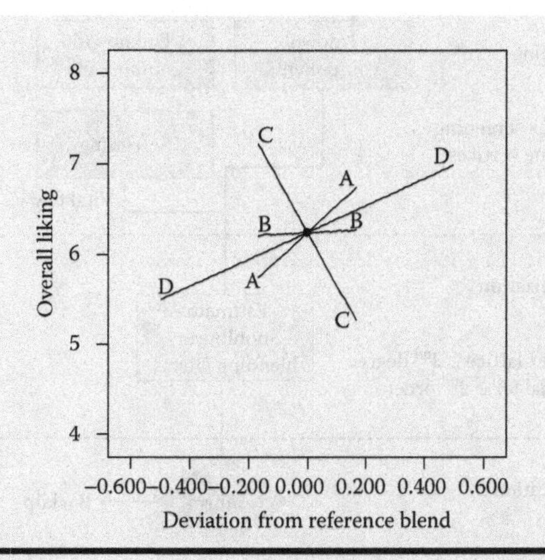

Figure 11.5 Trace plot for the chocolate-covered cherry experiment.

However, the mixture design produced an unexpected result—consumers liked getting some cherries flipped upside down in any given box—hence the sharp upward slope to component A. Who knew?

My momma always said, "Life is like a box of chocolates. You never know what you're gonna get."

—Forrest Gump (Paramount Pictures, 1994)

Strategy for Formulation Experimentation

In Chapter 8 we devoted our attention to mixture-screening designs, which we advise for experiments on 6 or more components. The flowchart shown in Figure 11.6 refines this recommendation by temporarily setting aside components you know to be important and screening only those whose impacts on the formulation remain relatively unknown.

After screening previously unknown components, the strategy calls for combining the vital few with the important ones you set aside earlier into an optimization design that models nonlinear blending effects. This requires a quadratic (second degree or second order) or higher (e.g., special cubic) design. The last strategic phase puts the predictive model to the test

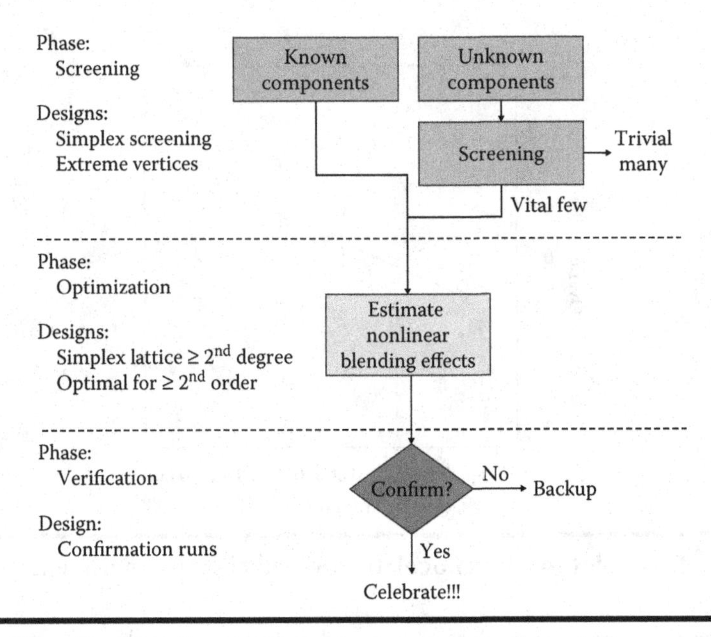

Figure 11.6 Flowchart for the strategy of mixture experimentation.

via confirmation runs; for example, those done in the coffee experiment detailed in Chapter 10, or more rigorous methods (see Martin Bezener's March 2015 webinar on "Practical Strategies for Model Verification" posted at www.statease.com/training/webinar.html).

PROPERLY SIZING YOUR MIXTURE DESIGN

Due to the collinearity of components that is inherent in mixture designs (i.e., once all but one of the ingredients is set, the last must make up the difference to the total amount), power calculations fall short. The reason is to ensure that, for a given variable that can naturally be expected in your blending process, the difference you wish to see will likely be detected. Instead, we advise you make use of the sizing tools that are detailed in the section on "Right-Sizing Designs via Fraction of Design Space (FDS) Plots" of *RSM Simplified, 2nd edition*, Chapter 12.

(*Continued*)

A designer knows he has achieved perfection not when there is nothing left to add, but when there is nothing left to take away.

—Antoine de Saint-Exupery (*Wind, Sand, and Stars*, 1939)
Read more at: https://www.brainyquote.com/quotes/topics/topic_design.html

This concludes our book on mixture design for optimal formulation. We hope you put these methods to good use for revising your recipe to the sweetest spot ever.

References

Adulteration. *Encyclopaedia Britannica*, 10th ed., 1902. www.1902encyclopedia.com/A/ADU/adulteration.html.

Anderson, M. Mixture design enhances the nectar from an exotic Amazonian fruit. *Stat-Teaser*, March 2005. Stat-Ease, Minneapolis, MN.

Anderson, M., and P. Whitcomb. Computer-aided design of experiments for formulations. *Modern Paint and Coatings*, June 1997a, pp. 38–39.

Anderson, M., and P. Whitcomb. How statisticians keep things simple. *RSM Simplified*. New York: Productivity, Press, 2005.

Anderson, M., and P. Whitcomb. Mixing it up with computer-aided design. *Today's Chemist at Work*, 1997b, p. 34.

Anderson, M., and P. Whitcomb. Mixture DOE uncovers formulations quicker. *Rubber & Plastics News*, October 21, 2002, pp. 16–18.

Anderson, M., and P. Whitcomb. *DOE Simplified: Practical Tools for Effective Experimentation*, 3rd ed. New York: Productivity Press, 2015.

Anderson, M., and P. Whitcomb. Practical aspects for designing statistically optimal experiments. *Journal of Statistical Science and Application*, 2014, 2, 85–92.

Anderson, M., and P. Whitcomb. *RSM Simplified: Optimizing Processes Using Response Surface Methods for Design of Experiments*, 2nd ed. New York: Productivity Press, 2016.

Ansmann, A. et al. Personal care formulations. Chapter 7 of the *Handbook of Detergents: Part D: Formulation*, M. S. Showell (Ed.). Boca Raton, FL: CRC Press, 2005, p. 248.

Bezener, M., and H. Anderson. Brewing the perfect pot of office coffee. *Stat-Teaser*, September 2016. Stat-Ease, Minneapolis, MN. www.statease.com/publications/newsletter/stat-teaser-09-16.

Blackledge, R. Finding the sweet spot of your cricket bat. Live and Breathe Cricket blog, December 14, 2012. http://liveandbreathecricket.blogspot.com/2012/12/finding-sweet-spot.html.

Box, G. E. P., and N. R. Draper. *Response Surfaces, Mixtures, and Ridge Analyses*, 2nd ed. New York: John Wiley & Sons, 2007.

Box, G. E. P., and H. L. Lucas. Design of experiments in nonlinear situations. *Biometrika*, 1959, 46, 77–90.

Cornell, J. *Experiments with Mixtures*, 3rd ed. New York: John Wiley & Sons, 2002.

Cornell, J., and G. Piepel. *Methods for Designing and Analyzing Mixture Experiments*, notes from a short course presented at the Fall Technical Conference by the Chemical & Process Industries Divisions (CPID) and the Statistics Division of the American Society for Quality (ASQ), and by the Section on Physical and Engineering Sciences (SPES) and the Section on Quality & Productivity (Q&P) of the American Statistical Association (ASA), October 8, 2008, Phoenix, AZ.

Cox, D. R. A note on polynomial response functions for mixtures. *Biometrika*, 1971, 58, 155–159.

Crosier, R. B. Mixture experiments: Geometry and pseudocomponents. *Technometrics*, 1984, 26(3), 209–216.

Daniel, C., and E. L. Lehmann. Henry Scheffe 1907–1977. *The Annals of Statistics*, 1979, 7(6), 1149–1161.

Del Vecchio, R. J. *Design of Experiments*. Hanser/Gardner, Cincinnati, OH, 1997, pp. 100–101.

DeVries, P. *Let Me Count the Ways*. Boston, MA: Little Brown, 1965.

Dick, P. K. *Valis (Valis Trilogy #1)*. Santa Ana, CA, 2011. www.amazon.com/VALIS-Valis-Trilogy-Philip-Dick/dp/0547572417.

Draper, N. R., and I. Guttman. Rationalization of the "alphabetic-optimal" and "variance plus bias" approaches to experimental design. Technical Report 841, 1988, Department of Statistics, University of Wisconsin.

Draper, N. R., P. Prescott, S. M. Lewis, A. M. Dean, P. W. M. John, and M. G. Tuck. Mixture designs for four components in orthogonal blocks. *Technometrics*, 1993, 35(3), 268–276.

Dryden, J. *All for Love; Or, The World Well Lost: A Tragedy*. 1678. www.bartleby.com/18/1/.

Goos, P., and A. N. Donev. The D-optimal design of blocked and split-plot experiments with mixture components, Research Report 0303, Departement Toegepaste Economische Wetenschappen, Katholieke Universiteit Leuven, January 2003, p. 1. https://lirias.kuleuven.be/bitstream/123456789/118367/1/OR_0303.pdf.

Guidance for Industry. Q8(R2) Pharmaceutical development. U.S. FDA, November 2009, p. 9.

Hensley, C. Design of experiments helps reduce time to remove aerospace coatings. *Aerospace Engineering & Manufacturing Technology Update*, 2008, pp. 21–23. http://articles.sae.org/2916/.

Humphrey, J. W., J. P. Oleson, and A. N. Sherwood. *Greek and Roman Technology: A Sourcebook*. Routledge, New York, 1998.

Kalichevsky, V. A. *The Amazing Petroleum Industry*. New York: Reinhold, 1943, p. 7.

Koons, G. F., and M. H. Wilt. Design and analysis of an acrylonitrile—butadiene—styrene (ABS) pipe compound experiment. *Computer Applications in Applied Polymer Science*, Chapter 27. Washington, DC: American Chemical Society, 1982, pp. 439–448.

Kowalski, S., J. A. Cornell, and G. G. Vining. A new model and class of designs for mixture experiments with process variables. *Communication in Statistics: Theory and Methods*, 2000, 29(9–10), 2255–2280.

Kowalski, S., J. A. Cornell, and G. G. Vining. Split-plot designs and estimation methods for mixture experiments with process variables. *Technometrics*, 2002, 44(1), 72.

Kris-Etherton, P., R. H. Eckel, B. V. Howeard, et al. Lyon Diet Heart Study: Benefits of a Mediterranean-lifestyle, National Cholesterol Education Program/American Heart Association Step I dietary pattern on cardiovascular disease. *Circulation*, 2001, 103, 1823–1825.

Lawson, J., and C. Willden. Mixture experiments in R using mixexp. *Journal of Statistical Software*, 2016, 72, 1–20.

Lewis, S. M., A. M. Dean, N. R. Draper, and P. Prescott. Mixture designs for q components in orthogonal blocks. *Journal of the Royal Statistical Society. Series B (Methodological)*, 1994, 56(3), 457–467.

McLean, R. A., and V. L. Anderson. Extreme vertices design of mixture experiments. *Technometrics*, 1966, 8(3), 449.

Meadows, S. L., C. Gennings, W. H. Carter Jr., and D. S. Bae. Experimental designs for mixtures of chemicals along fixed ratio rays–classic methodology for detecting and characterizing departures from additivity, isobolograms. *Environmental Health Perspectives*, 2002, 110(Suppl 6), 979.

Myers, R. H., D. C. Montgomery, and C. M. Anderson-Cook. *Response Surface Methodology, Process and Product Optimization Using Designed Experiments*, 3rd ed. New York: John Wiley & Sons, 2009.

Piepel, G. F. Measuring component effects in constrained mixture experiments. *Technometrics*, 1982, 25, 97–105.

Piepel, G. F. Programs for generating extreme vertices and centroids of linearly constrained experimental regions. *Journal of Quality Technology*, 1988, 20(2), 125–139.

Piepel, G. F., and J. A. Cornell. Designs for mixture-amount experiments. *Journal of Quality Technology*, 1987, 19(1), 11–28.

Roesler, R. R. How to bake the perfect cake. *Paint & Coatings Industry*, 2004.

Roudsari, S. F., R. Dhib, and F. Ein-Mozaffari. Mixing effect on emulsion polymerization in a batch reactor. *Polymer Engineering and Science*, 2014, 55(4), 945–956.

Ruhlman, M. *Ratio: The Simple Codes behind the Craft of Everyday Cooking*, 2009, Scribner, New York, p. ix.

Sahrmann, H. F., G. F. Piepel, and J. A. Cornell. In search of the optimum Harvey Wallbanger recipe via mixture experiment techniques. *American Statistician*, 1987, 41, 190–194.

Scheffé, H. Experiments with mixtures. *Journal of Royal Statistical Society*, 1958, B20, 344–360.

Scheffe, H. Simplex-centroid design for experiments with mixtures. *Journal of the Royal Statistical Society, Series B (Methodological)*, 1963, 25(2), 235–263.

Simon, H. *Administrative Behavior: A Study of Decision-Making Processes in Administrative Organization*, 4th ed. New York: The Free Press, 1997, p. xxix.

Singh, S. K., J. Dodge, M. J. Durrani, and M. A. Khan. Optimization and characterization of controlled release pellets coated with an experimental latex: I. Anionic drug. *International Journal of Pharmaceutics*, 1995, 125(2), 243–255.

Smith, W. *Experimental Design for Formulation*. Philadelphia, PA: ASA-SIAM Series on Statistics and Applied Probability, 2005.

Snee, R. D. Experimental designs for mixture systems with multicomponent constraints. *Communications in Statistics*, 1979, 303–306.

Snee, R. D., R. W. Hoerl, and G. Bucci. A statistical engineering approach to mixture experiments with process variables. *Quality Engineering*, 2016, 28(3), 263–279.

Snee, R. D., and D. W. Marquardt. Extreme vertices designs for linear mixture models. *Technometrics*, 1974, 16(4), 399–408.

Snee, R. D., and D. W. Marquardt. Screening concepts and designs for experiments with mixtures. *Technometrics*, 1976, 18(1), 24.

Spagnola, L., J. Klang, M. Gupta, and X. Drujon. New advances in UV-curable soft-touch coatings. *Coatings World*, October 8, 2016. https://www.coatingsworld.com/issues/2016-08-01/view_features/new-advances-in-uv-curablesoft-touch-coatings/.

Suzuki, J. *A History of Mathematics*. Upper Saddle River, NJ: Prentice Hall, 2002, p. 731.

Thomson, J. R., and J. R. Thomson. Taste Senior Editor. Which has more caffeine, light or dark roast? Here's the truth. *Huffington Post*, June 27, 2016. http://www.huffingtonpost.com/entry/coffee-roast-caffeine_us_576c8725e4b017b379f57ff8.

Vieira, M. C., and C. L. M. Silva. Optimization of a cupuacu (Theobroma grandiflorum) nectar formulation. *Journal of Food Process Engineering*, 2004, 27, 181–196.

Vojnovic, D., B. Campisi, A. Mattei, and L. Favretto. Experimental mixture design to ameliorate the sensory quality evaluation of extra virgin oils. *Chemometrics and Intelligent Laboratory Systems*, 1995, 27, 205–210.

Whitcomb, P. *Mixture Design for Optimal Formulations Workshop*. Section 1. Slide 3. Stat-Ease, Minneapolis, MI, 2009.

Whitcomb, P. J., and M. J. Anderson. Using DOE with Tolerance Intervals to Verify Specifications. *11th Annual Meeting of the European Network for Business and Industrial Statistics (ENBIS)*, University of Coimbra, Coimbra, Portugal, September 2011, pp. 4–8. www.statease.com/pubs/using_DOE_with_tolerance_intervals.pdf.

About the Software

To make mixture design for optimal formulation easy, this book is augmented with a fully functional, time-limited, version of a commercially available computer program from Stat-Ease, Inc.—called Design-Expert® software. Download this Windows-based computational tool, as well as enlightening tutorials, from www.statease.com/formulation-simplified.html. There you will also find files of data for most of the exercises in the book. The datasets are named to be easily cross-referenced with corresponding material in the book. Also, if you work through all the problems (the only way to a working knowledge of mixture design), check your answers by downloading them from the book's website.

You are encouraged to reproduce the results shown in the book and to explore further. The Stat-Ease software offers far more detail in statistical outputs and many more graphics than can be included in this book. You will find a great deal of information on program features and statistical background in the help system built into the software.

Technical support for the software can be obtained by contacting:

Stat-Ease, Inc.
2021 East Hennepin Ave, Suite 480
Minneapolis, MN 55413
Telephone: 612-378-9449
Fax: 612-378-2152
E-mail: support@statease.com
Website: www.statease.com

Index

Note: Page numbers followed by f and t refer to figures and tables respectively.

G

H

I

J

K

L

U

V

W